Learning and Teaching Geometry, K–12

1987 Yearbook

Mary Montgomery Lindquist
1987 Yearbook Editor
Columbus College

Albert P. Shulte
General Yearbook Editor
Oakland Schools, Pontiac, Michigan

**National Council of
Teachers of Mathematics**

Printed in the United States of America

Contents

Part 1: Perspectives

Part 2: A View of Problem Solving and Applications

Part 3: Activities in Focus

Part 4: A Look at Geometry and Other Mathematics

Part 5: Preparing Teachers

Preface

While I was working on this yearbook, my husband gave me an addition to my collection of old mathematics books. This small book, *Practical Geometry,* was written in 1690 by the French mathematician Sebastian Le Clerc. It is illustrated with eighty copperplates (see figure below) among which, "besides the several Geometric Figures, are contained many Examples of Landskips, Pieces of Architecture, Perspective, Draughts of Figures, Ruins, &c." Charmed by these illustrations, I was also struck with the clarity of the book's purpose:

> Geometry is divided into speculative and practical.
>
> The former is a science that teaches the mind how to form ideas of, and demonstrate the truth of geometrical propositions.
>
> The latter, or practical Geometry, conducts the hand in working. (P. A2)

This yearbook sets forth the position that today there is a great body of knowledge that should be called on when decisions are being made about the geometry curriculum. In particular, there is research about how students learn and what they are not learning. In addition, there is a wealth of suggestions about the teaching of particular topics in geometry, both the "speculative" and the "practical." There are ways to integrate geometry into other mathematical areas as well as ways to use geometry to enhance these areas. Geometry is a natural topic to encourage problem solving, and it has many new applications that broaden the meaning of "practical." The yearbook gives a sampler of this knowledge including that on which you can speculate and that which you can use in a practical way.

Whenever one learns and works with capable people, the experience is a pleasant one; and so it has been in editing this yearbook. The authors, to whom the main credit is due, were professional in every way, from their thoughtful writing to their prompt responses at each stage of producing the book. Over and over, authors expressed their appreciation for the careful editing done by the staff at Reston. Athough I have already shared these comments with the editorial group at Reston, they deserve public recognition for a job well done. The advisory board brought to the project a diversity of views, backgrounds, and talents. Without them, the yearbook

could not have taken shape. My thanks and appreciation to each member
of the advisory board:

Donald W. Crowe	University of Wisconsin—Madison
Douglas A. Grouws	University of Missouri—Columbia
George A. Milauskas	Barrington High School, Illinois
Paula J. Raabe	Saint Paul Lutheran School, Illinois
Albert P. Shulte	Oakland Schools, Michigan.

MARY MONTGOMERY LINDQUIST
1987 Yearbook Editor

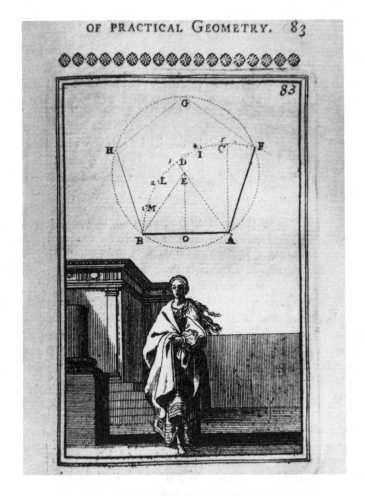

This figure appears in *Practical Geometry,* written in 1690
by the French mathematician Sebastian Le Clerc (London:
John Bowles and Carrington Bowles, 1768).

The van Hiele Model of the Development of Geometric Thought

Mary L. Crowley

HAVE you ever had students who could recognize a square but not define it? Have you ever noticed that some students do not understand that a square is a rectangle? Have you ever had students who complain about having to prove something they already "know"? According to two Dutch educators, Dina van Hiele-Geldof and her husband, Pierre Marie van Hiele, behaviors such as these reflect a student's level of geometric maturity. Have you ever wondered how to help your students achieve a more sophisticated level of geometric thinking? The van Hiele model of geometric thought can be used to guide instruction as well as assess student abilities. This article presents an overview of the model and discusses its classroom implications.

The van Hiele model of geometric thought emerged from the doctoral works of Dina van Hiele-Geldof (1984a) and Pierre van Hiele (1984b), which were completed simultaneously at the University of Utrecht. Since Dina died shortly after finishing her dissertation, it was Pierre who clarified, amended, and advanced the theory. With the exception of the Soviet Union, where the geometry curriculum was revised in the 1960s to conform to the van Hiele model, the work was slow in gaining international attention. It was not until the 1970s that a North American, Izaak Wirszup (1976), began to write and speak about the model. At about the same time, Hans Freudenthal, the van Hieles' professor from Utrecht, called attention to their works in his titanic book, *Mathematics as an Educational Task* (1973). During the past decade there has been increased North American interest in the van Hieles' contributions. This has been particularly enhanced by the 1984 translations into English of some of the major works of the couple (Geddes, Fuys, and Tischler 1984).

The model consists of five levels of understanding. The levels, labeled "visualization," "analysis," "informal deduction," "formal deduction," and "rigor" (Shaughnessy and Burger 1985, p. 420) describe characteristics of the thinking process. Assisted by appropriate instructional experiences, the model asserts that the learner moves sequentially from the initial, or basic, level (visualization), where space is simply observed—the properties of figures are not explicitly recognized, through the sequence listed above to the highest level (rigor), which is concerned with formal abstract aspects of deduction. Few students are exposed to, or reach, the latter level. A synopsis of the levels is presented below.

The Model

Level 0[1] (Basic Level): Visualization

At this initial stage, students are aware of space only as something that exists around them. Geometric concepts are viewed as total entities rather than as having components or attributes. Geometric figures, for example, are recognized by their shape as a whole, that is, by their physical appearance, not by their parts or properties. A person functioning at this level can learn geometric vocabulary, can identify specified shapes, and given a figure, can reproduce it. For example, given the diagrams in figure 1.1, a student at this level would be able to recognize that there are squares in (a) and rectangles in (b) because these are similar in shape to previously encountered squares and rectangles. Furthermore, given a geoboard or paper, the student could copy the shapes. A person at this stage, however, would not recognize that the figures have right angles or that opposite sides are parallel.

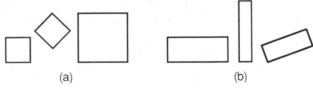

(a)   (b)

Fig. 1.1

Level 1: Analysis

At level 1, an analysis of geometric concepts begins. For example, through observation and experimentation students begin to discern the characteristics of figures. These emerging properties are then used to conceptualize classes of shapes. Thus figures are recognized as having parts and are recognized by their parts. Given a grid of parallelograms such as those in figure 1.2, students could, by "coloring" the equal angles, "establish" that the opposite angles of parallelograms are equal. After using several such examples, students could make generalizations for the class of parallelograms. Relationships between properties, however, cannot yet be explained by students at this level, interrelationships between figures are still not seen, and definitions are not yet understood.

---

1. Different numbering systems for the model may be encountered in the literature. The van Hieles themselves spoke of levels beginning with the basic level, or level 0, and ending with level 4.

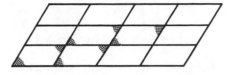

Fig. 1.2

Level 2: Informal Deduction

At this level, students can establish the interrelationships of properties both within figures (e.g., in a quadrilateral, opposite sides being parallel necessitates opposite angles being equal) and among figures (a square is a rectangle because it has all the properties of a rectangle). Thus they can deduce properties of a figure and recognize classes of figures. Class inclusion is understood. Definitions are meaningful. Informal arguments can be followed and given. The student at this level, however, does not comprehend the significance of deduction as a whole or the role of axioms. Empirically obtained results are often used in conjunction with deduction techniques. Formal proofs can be followed, but students do not see how the logical order could be altered nor do they see how to construct a proof starting from different or unfamiliar premises.

Level 3: Deduction

At this level, the significance of deduction as a way of establishing geometric theory within an axiomatic system is understood. The interrelationship and role of undefined terms, axioms, postulates, definitions, theorems, and proof is seen. A person at this level can construct, not just memorize, proofs; the possibility of developing a proof in more than one way is seen; the interaction of necessary and sufficient conditions is understood; distinctions between a statement and its converse can be made.

Level 4: Rigor

At this stage the learner can work in a variety of axiomatic systems, that is, non-Euclidean geometries can be studied, and different systems can be compared. Geometry is seen in the abstract.

This last level is the least developed in the original works and has received little attention from researchers. P. M. van Hiele has acknowledged that he is interested in the first three levels in particular (Alan Hoffer, personal communication, 25 February 1985). Since the majority of high school geometry courses are taught at level 3, it is not surprising that most research has also concentrated on lower levels. Perhaps as the van Hiele model is extended to other areas (it is being applied to economics and chemistry in Holland), this last level will achieve more prominence.

Properties of the Model

In addition to furnishing insights into the thinking that is specific to each level of geometric thought, the van Hieles identified some generalities that characterize the model. These properties are particularly significant to educators because they provide guidance for making instructional decisions.

1. *Sequential*. As with most developmental theories, a person must proceed through the levels in order. To function successfully at a particular level, a learner must have acquired the strategies of the preceding levels.

2. *Advancement*. Progress (or lack of it) from level to level depends more on the content and methods of instruction received than on age: No method of instruction allows a student to skip a level; some methods enhance progress, whereas others retard or even prevent movement between levels. Van Hiele points out that it is possible to teach "a skillful pupil abilities above his actual level, like one can train young children in the arithmetic of fractions without telling them what fractions mean, or older children in differentiating and integrating though they do no know what differential quotients and integrals are" (Freudenthal 1973, p. 25). Geometric examples include the memorization of an area formula or relationships like "a square is a rectangle." In situations like these, what has actually happened is that the subject matter has been reduced to a lower level and understanding has not occurred.

3.- *Intrinsic and extrinsic*. The inherent objects at one level become the objects of study at the next level. For example, at level 0 only the form of a figure is perceived. The figure is, of course, determined by its properties, but it is not until level 1 that the figure is analyzed and its components and properties are discovered.

4. *Linguistics*. "Each level has its own linguistic symbols and its own systems of relations connecting these symbols" (P. van Hiele 1984a, p. 246). Thus a relation that is "correct" at one level may be modified at another level. For example, a figure may have more than one name (class inclusion)—a square is also a rectangle (and a parallelogram!). A student at level 1 does not conceptualize that this kind of nesting can occur. This type of notion and its accompanying language, however, are fundamental at level 2.

5. *Mismatch*. If the student is at one level and instruction is at a different level, the desired learning and progress may not occur. In particular, if the teacher, instructional materials, content, vocabulary, and so on, are at a higher level than the learner, the student will not be able to follow the thought processes being used.

Phases of Learning

As was indicated above, the van Hieles assert that progress through the levels is more dependent on the instruction received than on age or maturation. Thus the method and organization of instruction, as well as the content and materials used, are important areas of pedagogical concern. To address these issues, the van Hieles proposed five sequential phases of learning: inquiry, directed orientation, explication, free orientation, and integration. They assert that instruction developed according to this sequence promotes the acquisition of a level (van Hiele-Geldof 1984b). Sample activities from level-2 work with the rhombus are used here to illustrate.

Phase 1: Inquiry/Information

At this initial stage, the teacher and students engage in conversation and activity about the objects of study for this level. Observations are made, questions are raised, and level-specific vocabulary is introduced (Hoffer 1983, p. 208). For example, the teacher asks students, "What is a rhombus? A square? A parallelogram? How are they alike? Different? Do you think a square could be a rhombus? Could a rhombus be a square? Why do you say that? . . ." The purpose of these activities is twofold: (1) the teacher learns what prior knowledge the students have about the topic, and (2) the students learn what direction further study will take.

Phase 2: Directed Orientation

The students explore the topic of study through materials that the teacher has carefully sequenced. These activities should gradually reveal to the students the structures characteristic of this level. Thus, much of the material will be short tasks designed to elicit specific responses. For example, the teacher might ask students to use a geoboard to construct a rhombus with equal diagonals, to construct another that is larger, to construct another that is smaller. Another activity would be to build a rhombus with four right angles, then three right angles, two right angles, one right angle. . . .

Phase 3: Explication

Building on their previous experiences, students express and exchange their emerging views about the structures that have been observed. Other than to assist students in using accurate and appropriate language, the teacher's role is minimal. It is during this phase that the level's system of relations begins to become apparent. Continuing the rhombus example, students would discuss with each other and the teacher what figures and properties emerged in the activities above.

Phase 4: Free Orientation

The student encounters more complex tasks—tasks with many steps, tasks that can be completed in several ways, and open-ended tasks. "They gain experience in finding their own way or resolving the tasks. By orienting themselves in the field of investigation, many relations between the objects of study become explicit to the students" (Hoffer 1983, p. 208). For example, students would complete an activity such as the following. "Fold a piece of paper in half, then in half again as shown here (fig. 1.3a). Try to imagine what kind of figure you would get if you cut off the corner made by the folds (fig. 1.3b). Justify your answer before you cut. What type(s) of figures do you get if you cut the corner at a 30° angle? At a 45° angle? Describe the angles at the point of intersection of the diagonals. The point of intersection is at what point on the diagonals? Why is the area of a rhombus described by one-half the product of the two diagonals?"

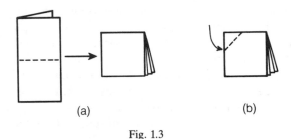

(a)   (b)

Fig. 1.3

Phase 5: Integration

The students review and summarize what they have learned with the goal of forming an overview of the new network of objects and relations. The teacher can assist in this synthesis "by furnishing global surveys" (van Hiele 1984a, p. 247) of what the students have learned. It is important, however, that these summaries not present anything new. The properties of the rhombus that have emerged would be summarized and their origins reviewed.

At the end of the fifth phase, students have attained a new level of thought. The new domain of thinking replaces the old, and students are ready to repeat the phases of learning at the next level.

Providing van Hiele–Based Experiences

Implicit in the van Hieles' writing is the notion that children should be presented with a wide variety of geometric experiences. Teachers in the early elementary years can provide basic-level exploratory experiences through cutouts, geoboards, paper folding, D-sticks, straws, grid work, tessellations, tangrams, and geometric puzzles. Middle school and junior high school experiences, roughly at levels 1 and 2, can include working with grids, collections of shapes, "property cards," "family trees," and "what's my name" games. The following pages provide examples of these and other types of activities appropriate for the first four van Hiele levels. Many of these ideas were culled from the descriptors of student behavior developed by the researchers at Brooklyn College (Geddes et al. 1985). Additional activities can be found in the articles by Burger (1985), Burger and Shaughnessy (1986), Hoffer (1981), Prevost (1985), and Shaughnessy and Burger (1985).

Basic Level (Visualization): Geometric shapes are recognized on the basis of their physical appearance as a whole.

Provide students opportunities—

1. to manipulate, color, fold, and construct geometric shapes

Pattern blocks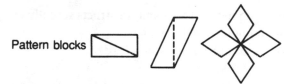

2. to identify a shape or geometric relation—
   - in a simple drawing
   - in a set of cutouts, pattern blocks, or other manipulatives (i.e., sort)
   - in a variety of orientations

   - involving physical objects in the classroom, the home, photographs, and other places

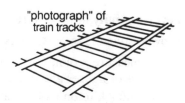
"photograph" of train tracks

   - within other shapes

parallel lines in a trapezoid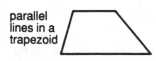

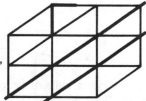
right angles, triangles, parallel lines, rectangles, etc.

3. to create shapes—
   - by copying figures on dot paper, grid paper, or tracing paper, by using geoboards and circular geoboards, or by tracing cutouts
   - by drawing figures
   - by constructing figures with sticks, straws, or pipe cleaners or by tiling with manipulatives, pattern blocks, and so on

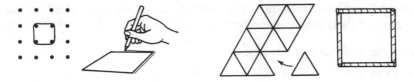

4. to describe geometric shapes and constructs verbally using appropriate standard and nonstandard language
   - a cube "looks like a block or a box"
   - "corners" for angles

5. to work on problems that can be solved by managing shapes, measuring, and counting
   - Find the area of a box top by tiling and counting.
   - Use two triangular shapes to make a rectangle; another triangle (tangrams).

Level 1 (Analysis): Form recedes and the properties of figures emerge.

Provide students opportunities—

1. to measure, color, fold, cut, model, and tile in order to identify properties of figures and other geometric relationships
   - fold a kite on a diagonal and examine the "fit."

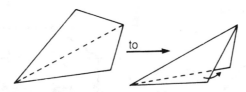

2. to describe a class of figures by its properties (charts, verbally, "property cards")

- "Without using a picture, how would you describe a [figure] to someone who has never seen one?"
- property cards

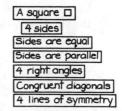

3. to compare shapes according to their characterizing properties
   - Note how a square and a rhombus are alike, are different in regard to angles, . . . in regard to sides.

4. to sort and resort shapes by single attributes
   - Sort cutouts of quadrilaterals by
     —number of parallel sides
     —number of right angles

5. to identify and draw a figure given an oral or written description of its properties
   - Teachers or students describe a figure verbally and ask for (all possible) figures with those properties.
   - "What's my name"—reveal clues (properties) one by one, pausing after each, until students can accurately identify the figure. This can be done on an overhead, piece of paper, property cards.

6. to identify a shape from visual clues
   - gradually reveal a shape, asking students to identify at each stage possible names for the shape.

7. to empirically derive (from studying many examples) "rules" and generalizations
   - From tiling and measuring many rectangles, students see that "$b \times h$" is a shortcut for adding the number of tiles.

8. to identify properties that can be used to characterize or contrast different classes of figures
   - Ask, "Opposite sides equal describes . . ."
   - Explore the relationship between diagonals and figures by joining two cardboard strips. A square is generated by the end points when . . . (the diagonals are congruent,

bisect each other, and meet at right angles). Change the angle and
the diagonals determine . . . (a rectangle). Noncongruent diagonals
generate . . .

9. to discover properties of unfamiliar classes of objects
   - From examples and nonexamples of trapezoids, determine the properties of trapezoids.

10. to encounter and use appropriate vocabulary and symbols

11. to solve geometric problems that require knowing properties of figures, geometric relationships, or insightful approaches
    - Without measuring, find the sum of the angles in a septagon. (Insightful students will "see" triangles, that is, relate this to known figures.)

Level 2 (Informal Deduction): A network of relations begins to form.

Provide students opportunities—

1. to study relationships developed at level 1, looking for inclusions and implications
   - Use property cards:

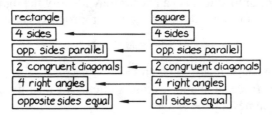

   - Working on a geoboard, change a quadrilateral to a trapezoid, trapezoid to parallelogram, parallelogram to rectangle. . . . What was required in each transformation?

2. to identify minimum sets of properties that describe a figure
   - Students could compete and check each other in this. Ask students how they would describe a figure to someone. Could they use fewer steps? Different steps?

3. to develop and use definitions
   - A square is . . .

4. to follow informal arguments

5. to present informal arguments (using diagrams, cut-out shapes, flow charts)
   - Ancestry mappings: Use cards and arrows to display the "origins" or

"family tree" of an idea—for example, "The exterior angle of a triangle equals the sum of the opposite interior angles."

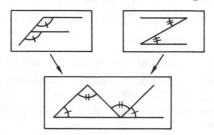

6. to follow deductive arguments, perhaps supplying a few "missing steps"
   - C is the center of the circle. Why is—
     a) $AC \cong BC$
     b) $\angle CAB \cong \angle CBA$
     c) $\triangle ACE \cong \triangle BCE$
     d) $AE \cong EB$

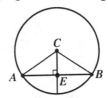

Note: Reasons other than the level-0 response, "It looks like . . .," must be given for this to be level 2.

7. to attempt to provide more than one approach or explanation
   - Define a parallelogram in two ways (i.e., "4 sides, opposite sides parallel" or "4 sides, opposite sides congruent").

8. to work with and discuss situations that highlight a statement and its converse
   - Write the converse of this statement: If a transversal intersects two parallel lines, then the interior angles on the same side of the transversal are supplementary. Which diagram correctly reflects the converse?

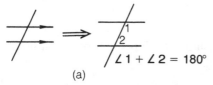

(a)

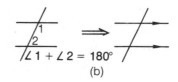

(b)

   - State the converse of the following statement and discuss its validity: "If it's raining, I'm wearing boots."

9. to solve problems where properties of figures and interrelationships are important
   - To construct the bisector of a line segment, sweep out two arcs of equal radii (as shown). Explain why the line through the points of intersection of

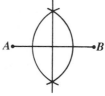

the arcs is the perpendicular bisector of the segment (i.e., use the properties of a rhombus).

Level 3 (Formal Deduction): The nature of deduction is understood. . . .

Provide students opportunities—

1. to identify what is given and what is to be proved in a problem
   - For the following problem, identify what is known and what is to be proved or shown. Do *NOT* complete the proof. "The perpendicular bisector of the base of an isosceles triangle passes through the vertex of the triangle."

2. to identify information implied by a figure or by given information
   - Figure *ABCD* is a parallelogram. Discuss what you know about this figure. Write a problem in "If . . . then . . ." form based on this figure.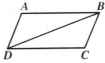

3. to demonstrate an understanding of the meaning of *undefined term, postulate, theorem, definition,* etc.,
   - Which of the following statements is a postulate, a theorem, a definition? Why?
     - *a)* Points that lie on the same line are called collinear.   (D)
     - *b)* Two points determine a line.   (P)
     - *c)* Every segment has exactly one midpoint.   (T)
     - *d)* The midpoint of a segment is said to bisect the segment.   (D)

4. to demonstrate an understanding of necessary and sufficient conditions
   - Write a *definition* of a square that begins
     - *a)* A square is a quadrilateral . . .
     - *b)* A square is a parallelogram . . .
     - *c)* A square is a rectangle . . .
     - *d)* A square is a rhombus . . .

5. to prove rigorously the relationships developed informally at level 2

6. to prove unfamiliar relationships

7. to compare different proofs of a theorem—for example, the Pythagorean theorem

8. to use a variety of techniques of proof—for example, synthetic, transformations, coordinates, vectors

9. to identify general strategies of proof
   - If a proof involves parallelism, try "saws," "ladders," or rotations of 180°.

10. to think about geometric thinking

- The following situations involve deductive or inductive thinking. Identify which type of thinking is involved and why.

  a) All goats have a beard. Sandy is a goat. Thus, Sandy has a beard.

  b) After measuring the angles in a number of quadrilaterals, Shelly announces, "The sum of the angles of a quadrilateral is 360°."

To be effective, activities like the preceding ones need to be placed in a context. The "Phases of Learning" section presents guidelines on the sequencing and delivery of geometric activities within a level. The "Properties of the Model" section also provides teaching advice. In particular, they suggest that geometric activities should not reduce the level of the geometric content, that whenever possible, materials should set the stage for further learning, and that language is important in the development and assessment of geometric understandings. These ideas are discussed further below.

Too often, geometry is taught in a mechanical way. Consider the fact that the sum of the angles of a triangle is 180°. Frequently, this fact is established by generalizing after measuring the angles of a few triangles, or worse, students are simply told the information. The latter tactic is an example of the reduction of the content level. Level-1 activities, such as the coloring of angles in a triangular grid (fig. 1.4) and the extension of that activity to identifying parallel lines in the grid, provide the student with a powerful means, both inductively and deductively, for understanding the concept. Insight into the reason why the angle sum is 180° is obtained from the grid work and concomitantly the groundwork is laid for the formal proof at level 3. An additional bonus with this particular development is that the same structure can be reused to demonstrate that the measure of the exterior angle of a triangle equals the sum of the measures of the two interior angles.

Fig. 1.4

Language, as well as thoughtfully chosen materials, plays an important role in the development of geometric thinking. It is essential that children talk about their linguistic associations for words and symbols and that they use that vocabulary. Such verbalization requires students to articulate consciously what might otherwise be vague and undeveloped ideas. It can also serve to reveal immature or misconceived ideas. Initially, children should be encouraged to express their geometric understandings in their own terms—"corner" for angle, "slanty" for the sides of a parallelogram, "straight" for parallel lines. Gradually, however, children should be introduced to standard terminology and encouraged to use it precisely. Just because children are using a word does not mean they attach the same meaning to it as their listener. For example, some children say that ⌊ is a right angle but that ⌋ is a left angle. Some say this shape (☐) is a square but when turned 45 degrees (◇), it no longer is one. In each example, children have incorrectly focused on orientation as a determining characteristic. (Perhaps they were shown figures only in "standard" position.) They are interpreting the terms *right angle* and *square* to have a narrow meaning. Children who operate with notions like these are limiting their development. Through conversations, teachers can uncover misconceptions and incomplete notions as well as build on correct perceptions.

The teacher's use of language is also important. For example, in work on level 1, terms such as *all, some, always, never, sometimes* should be modeled and encouraged. Level-2 phrases include "it follows that . . ." and "if . . ., then. . . ." Level 3 would use and stress the meanings of *axiom, postulate, theorem, converse, necessary and sufficient,* and so on.

Teacher questioning is a crucial factor in directing student thinking. At all levels, asking children how they "know" is important. It is not enough, for example, for students at level 2 to be asked what is the sum of the angles of a pentagon. They should be challenged to explain why and to think about their explanation—could it be shown another way? "Raising appropriate questions, allowing a sufficient response-time and discussing the quality of the answers are methods that take into account level of thinking" (Geddes et al. 1985, p. 242).

For growth to occur, it is essential to match instruction with the student's level. Thus teachers must learn to identify students' levels of geometric thought. Because the nature of a student's geometric explanations reflects that student's level of thinking, questioning is an important assessment tool. As an example, consider responses to the questions "What type of figure is this? ☐ How do you know?" Students at each level are able to respond "rectangle" to the first question. (If a student does not know how to name the figure, he or she is not at level 0 for rectangles.) Examples of level-specific responses to the second question are given below. In parentheses is a brief explanation of why the statement reflects the assigned level.

Level 0: "It looks like one!" or "Because it looks like a door." (The answer is based on a visual model.)

Level 1: "Four sides, closed, two long sides, two shorter sides, opposite sides parallel, four right angles . . ." (Properties are listed; redundancies are not seen.)

Level 2: "It is a parallelogram with right angles." (The student attempts to give a minimum number of properties. If queried, she would indicate that she knows it is redundant in this example to say that opposite sides are congruent.)

Level 3: "This can be proved if I know this figure is a parallelogram and that one angle is a right angle." (The student seeks to prove the fact deductively.)

Additional examples of level-specific student behaviors can be found in *An Investigation of the van Hiele Model of Thinking in Geometry among Adolescents* (Geddes et al. 1985, pp. 62–78) and in "Characterizing the van Hiele Levels of Development in Geometry" (Burger and Shaughnessy 1986, pp. 41–45).

The model of geometric thought and the phases of learning developed by the van Hieles propose a means for identifying a student's level of geometric maturity and suggest ways to help students to progress through the levels. Instruction rather than maturation is highlighted as the most significant factor contributing to this development. Research has supported the accuracy of the model for assessing student understandings of geometry (Burger 1985; Burger and Shaughnessy 1986; Geddes et al. 1982; Geddes, Fuys, and Tischler 1985; Mayberry 1981; Shaughnessy and Burger 1985; Usiskin 1982). It has also shown that materials and methodology can be designed to match levels and to promote growth through the levels (Burger 1985; Burger and Shaughnessy 1986; Geddes et al. 1982; Geddes, Fuys, and Tischler 1985; Shaughnessy and Burger 1985). The need now is for classroom teachers and researchers to refine the phases of learning, develop van Hiele–based materials, and implement those materials and philosophies in the classroom setting. Geometric thinking can be accessible to everyone.

REFERENCES

Burger, William F. "Geometry." *Arithmetic Teacher* 32 (February 1985): 52–56.

Burger, William F., and J. Michael Shaughnessy. "Characterizing the van Hiele Levels of Development in Geometry." *Journal for Research in Mathematics Education* 17 (January 1986): 31–48.

Freudenthal, Hans. *Mathematics as an Educational Task.* Dordrecht, Netherlands: D. Reidel, 1973.

Geddes, Dorothy, David Fuys, C. James Lovett, and Rosamond Tischler. "An Investigation of the van Hiele Model of Thinking in Geometry among Adolescents." Paper presented at the annual meeting of the American Educational Research Association, New York, March 1982.

Geddes, Dorothy, David Fuys, and Rosamond Tischler. "An Investigation of the van Hiele Model of Thinking in Geometry among Adolescents." Final report, Research in Science Education (RISE) Program of the National Science Foundation, Grant No. SED 7920640. Washington, D.C.: NSF, 1985.

Hoffer, Alan. "Geometry Is More Than Proof." *Mathematics Teacher* 74 (January 1981): 11–18.

———. "Van Hiele–based Research." In *Acquisition of Mathematical Concepts and Processes,* edited by R. Lesh and M. Landau. New York: Academic Press, 1983.

Mayberry, Joanne W. "An Investigation in the Van Hiele Levels of Geometric Thought in Undergraduate Preservice Teachers" (Doctoral dissertation, University of Georgia, 1979). *Dissertation Abstracts International* 42 (1981): 2008A. (University Microfilms No. 80-23078)

Prevost, Fernand J. "Geometry in the Junior High School." *Mathematics Teacher* 78 (September 1985): 411–18.

Shaughnessy, J. Michael, and William F. Burger. "Spadework Prior to Deduction in Geometry." *Mathematics Teacher* 78 (September 1985): 419–28.

Usiskin, Zalman. "Van Hiele Levels and Achievement in Secondary School Geometry." Final report, Cognitive Development and Achievement in Secondary School Geometry Project. Chicago: University of Chicago, 1982.

van Hiele, Pierre M. "A Child's Thought and Geometry." In *English Translation of Selected Writings of Dina van Hiele-Geldof and Pierre M. van Hiele,* edited by Dorothy Geddes, David Fuys, and Rosamond Tischler as part of the research project "An Investigation of the van Hiele Model of Thinking in Geometry among Adolescents," Research in Science Education (RISE) Program of the National Science Foundation, Grant No. SED 7920640. Washington, D.C.: NSF, 1984a. (Original work published in 1959.)

———. "English Summary by Pierre Marie van Hiele of the Problem of Insight in Connection with School Children's Insight into the Subject Matter of Geometry." In *English Translation of Selected Writings of Dina van Hiele-Geldof and Pierre M. van Hiele,* edited by Dorothy Geddes, David Fuys, and Rosamond Tischler as part of the research project "An Investigation of the van Hiele Model of Thinking in Geometry among Adolescents," Research in Science Education (RISE) Program of the National Science Foundation, Grant No. SED 7920640. Washington, D.C.: NSF, 1984b. (Original work published in 1957.)

van Hiele-Geldof, Dina. "Dissertation of Dina van Hiele-Geldof Entitled: The Didactic of Geometry in the Lowest Class of Secondary School." In *English Translation of Selected Writings of Dina van Hiele-Geldof and Pierre M. van Hiele,* edited by Dorothy Geddes, David Fuys, and Rosamond Tischler as part of the research project "An Investigation of the van Hiele Model of Thinking in Geometry among Adolescents," Research in Science Education (RISE) Program of the National Science Foundation, Grant No. SED 7920640. Washington, D.C.: NSF, 1984a. (Original work published in 1957.)

———. "Last Article Written by Dina van Hiele-Geldof entitled: Didactics of Geometry as Learning Process for Adults." In *English Translation of Selected Writings of Dina van Hiele-Geldof and Pierre M. van Hiele,* edited by Dorothy Geddes, David Fuys, and Rosamond Tischler as part of the research project "An Investigation of the van Hiele Model of Thinking in Geometry among Adolescents," Research in Science Education (RISE) Program of the National Science Foundation, Grant No. SED 7920640. Washington, D.C.: NSF, 1984b. (Original work published in 1958.)

Wirszup, Izaak. "Breakthroughs in the Psychology of Learning and Teaching Geometry." In *Space and Geometry: Papers from a Research Workshop,* edited by J. Martin. Columbus, Ohio: ERIC/SMEAC, 1976.

Resolving the Continuing Dilemmas in School Geometry

Zalman Usiskin

$\mathbf{A}$LMOST all writings on school geometry are derived from two major problems: the poor performance of students and an outdated curriculum. These problems have been with us for some time. In 1969, Carl Allendoerfer wrote:

> The mathematical curriculum in our elementary and secondary schools faces a serious dilemma when it comes to geometry. It is easy to find fault with the traditional course in geometry, but sound advice on how to remedy these difficulties is hard to come by . . . . curricular reform groups at home and abroad have tackled the problem, but with singular lack of success or agreement. . . . We are, therefore, under pressure to "do something" about geometry; but what shall we do? (P. 165)

Why is it that our efforts to solve these problems have resulted in a "singular lack of success or agreement"? The thesis of this article is that two fundamental dilemmas confound attempts to reach viable solutions. The purpose of this article is to discuss these problems, identify the related dilemmas, and then provide suggestions for their resolution.

The Performance Problem

Except for the knowledge of shapes (something many children begin to learn even before entering first grade), the geometry knowledge of students at the end of elementary school is spotty and rather minimal. For example, on the 1982 National Assessment, fewer than 10% of 13-year-olds could find the measure of the third angle of a triangle given the measures of the other two angles. A more difficult calculation, obtaining the length of the hypotenuse given the lengths of the other two sides, was correctly performed by 20% of 13-year-olds (Carpenter et al. 1983).

These results indicate that the Pythagorean theorem is taught to more students in this age group than the triangle-sum theorem and illustrate the fundamental connection between curriculum and performance. If a topic is not taught, it is not learned. In grades K–8, although all leaders and almost all teachers realize that geometry is important enough to warrant a place at

17

all levels, there is no consistent agreement regarding the content, sequence, or timing of geometry to be taught. That is, there is no standard curriculum for elementary school geometry comparable to the curriculum that exists for arithmetic. As a result, most teachers cannot expect their students to have previously amassed anything more than the most rudimentary knowledge of geometry. Even great teachers can take students with such poor backgrounds only so far.

At the senior high school level in the United States, only about 52% of 17-year-olds report having taken a semester of geometry (NAEP 1983). Even these students, rather universally better than those who opt not to take geometry, enter geometry with sparse knowledge. Consider the question in figure 2.1. This item was part of a test given to 99 geometry classes in 13 senior high schools (constituting all geometry classes in those schools) in 5 states at the beginning of the 1980–81 school year (Usiskin 1982). Only 63% of entering geometry students correctly answered this item. A similar item involving squares was correctly answered by 90% of the students when the square's sides were horizontal or vertical, but by only 80% when the sides were tilted.

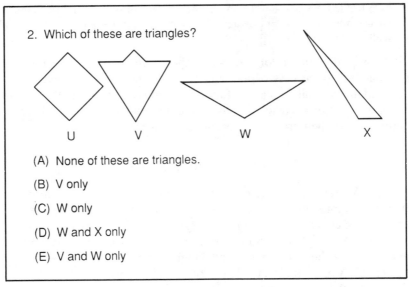

2. Which of these are triangles?

U   V   W   X

(A) None of these are triangles.

(B) V only

(C) W only

(D) W and X only

(E) V and W only

Fig. 2.1. Item 2 from the Van Hiele Geometry Test, developed by the CDASSG project in the Department of Education, University of Chicago. © 1980 by The University of Chicago. Used with permission.

Since students who do not take geometry would be likely to perform more poorly, we can conclude that the 48% of all students who do not take secondary school geometry know almost no geometry when they finish high school.

The ability to write proofs is considered one of the preeminent goals of high school geometry. Of the 99 classes in the study mentioned above, 85 studied proof long enough for the teachers to believe that a test on proof at the end of the year would be fair. At the end of the year in these classes, *28% of the students could not do the* easiest *triangle congruence proof.* Only 31% of the students were judged competent in proof (Senk 1985). There were high correlations between the knowledge of simple geometry facts at the beginning of the year and the performance on proof at the end of the year.

The Performance Dilemma

The lack of success that characterizes so many students' experiences in geometry discourages other students from taking geometry, encourages counselors to dissuade non-college-bound students from its study, keeps elementary school teachers from wanting to take geometry in college or teach it to their students, and perpetuates the cycle of poor performance. This chain of events constitutes the performance dilemma, a dilemma of the chicken-and-egg variety. To increase the performance of students, we must enlarge the pool of people who want to study geometry. To enlarge that pool, we must have a greater number of students who perform well in their study of geometry.

The discussion above suggests a number of steps to be taken:

1. *Specify an elementary school geometry curriculum by grade level.* It is difficult to raise standards of performance nationwide in elementary school geometry without some sort of nationally agreed-on core curriculum. The detail will have to be as much as the detail in which arithmetic is specified. Furthermore, standardized tests will have to be created to cover this new content. It will probably be best to begin with some core content and a variety of options. There is too much to learn to tolerate a situation in which later courses cannot assume that students have had some geometry in earlier courses. For example, tenth-grade geometry books should treat the measuring of angles and calculations of areas of simple plane figures as review. If senior high school teachers act as if this is new content, then there is no reason for teachers in earlier grades to teach it.

2. *Do not keep students from studying geometry merely because they are poor at arithmetic or algebra.* It is ironical that we test students on arithmetic to decide whether or not they can study algebra. Then we require a certain grade in algebra before a student can study secondary school geometry. It is as if we say, "Because you are not so good at basketball, we won't let you bowl either." Too often we find out what the student *cannot* do and then

give the student no opportunity to show what he or she *can* do. Every teacher knows that there are students who fare much better in geometry than in algebra or arithmetic. Students who do not possess much ability to multiply or divide may even be gifted visually.

3. *Require a significant amount of competence in geometry from* all *students.* Students leave elementary school not knowing enough geometry to succeed in high school (Usiskin 1982). As a result, only a minority of the half of students who take geometry achieve success. After the one-year geometry course, geometry is relatively ignored, and so even what is learned tends to be forgotten. Those who do not take geometry in high school are even worse off.

Reading requires a perception of geometric symbols (typefaces, scripts, printing). Virtually anyone who can read would seem to possess the perceptual ability necessary to understand geometry in any of its forms. At the high school level, girls and boys have equal ability to learn geometry and do geometric proofs (Senk and Usiskin 1983).

4. *Require that all prospective teachers of mathematics study geometry at the college level.* Today's elementary school teachers have not, for the most part, taken any geometry since high school, and some not even then. Even teachers certified in mathematics may not possess enough background to understand the issues, for college mathematics departments have cut back their teaching of geometry. Many new teachers have never studied geometry in three dimensions, may never have encountered a non-Euclidean geometry, and may not have dealt with transformations or vectors.

In the past few years, many states and schools have broadened and strengthened requirements for all mathematics students. It is also necessary to do the same for all mathematics teachers. We cannot expect elementary school teachers to teach a broadened curriculum in mathematics if, at the college level, they have only taken a course in the teaching of arithmetic. We cannot tolerate a situation in which some high school teachers teach geometry and others avoid it like the plague.

The Curriculum Problem

Implementing the four suggestions given above is not as easy as it might seem. There is a lack of agreement regarding not just the details but even the nature of the geometry that should be taught from elementary school through college.

The general public might find it difficult to believe that even at the elementary school level, there are few geometry concepts that are taught to everyone worldwide. When the French mathematician Jean Dieudonné intoned, "Euclid must go!" in 1959 (OECD 1960), he paved the way for a geometry curriculum in the French-speaking areas of Europe that bears little resemblance to curricula in other countries. As one consequence, in the amassing of items to test eighth-grade students in 1981–82 from 23 countries in the Second International Study of Mathematics Achievement, only 2 or 3 geometry items out of 40 proposed were agreed on by all countries (Kenneth Travers, personal communication, 27 September 1985). As a result, many compromises had to be made, and every country had to test with some items that were totally outside the experiences of its students. For example, an item from the eighth-grade geometry test that resulted from the French influence is given in figure 2.2.

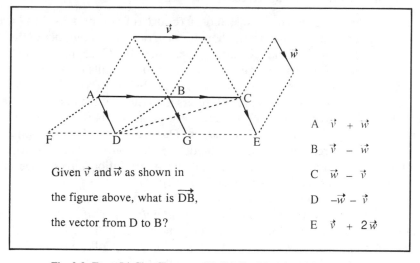

Fig. 2.2. From Li-Chu Chang and Judith Ruzicka, *Second International Mathematics Study—United States Technical Report I: Item Level Achievement Data, Eighth and Twelfth Grades* (Champaign, Ill.: Stipes Publishing Co., 1985).

At the secondary level, the curricular problem is more complicated. An "approach" to geometry means more than just content; it signals a logical framework in which the content is to be put or a type of proof is to be used. Allendoerfer (1969) identified three approaches to geometry—synthetic, analytic, and vector—and accurately predicted that "the discussion of the relative advantages of these three approaches is not likely to result in any conclusion that is widely acceptable" (p. 166).

The 1973 NCTM Yearbook on geometry identified the following approaches: conventional, affine, transformation, coordinate, vector, and eclectic. Each approach was discussed in an article, exemplifying what might be termed the "equal time" method of resolving the curricular problem. Similar discussions appeared at NCTM conventions and in grouped articles (e.g., see Fehr, Eccles, and Meserve [1972]), but they seldom served to clarify anything. They only confounded the dilemma by raising alternatives, and seldom was there time or space to discuss the subtle issues. Allendoerfer's prediction has held up over time.

At the college level, the debate is not merely on what to teach but on whether there should be any geometry courses at all. One mathematician has put it succinctly (Grünbaum 1981):

> [W]hy do I see geometry in colleges and universities in a very gloomy light? The simple answer is that we teach very little geometry, and that what we do teach is rather misleading. (P. 235)

This curricular problem is reflected in reports of committees that have made recommendations for the mathematics curriculum. A number of these reports have ignored geometry: among them the NACOME report (1975), the report of an NIE-sponsored conference on basic skills in Euclid, Ohio (NIE 1975), and Science and Mathematics in the Schools (NAS-NAE 1982). NCTM's *An Agenda for Action* (1980) refers the reader to the famous NCSM list of ten basic skills (1977, 1978), but there geometry is treated only in a single paragraph. In *School Mathematics: Options for the 1990s* (Romberg 1984), only one recommendation is made for the course, that "the topics should be unified and integrated so that the interrelationships of algebra, geometry, and applications are made" (p. 12).

Even when the population is narrowed, the recommendations tend to be brief and without careful argument. In *Academic Preparation for College,* the College Board (1983) covers geometry in nine sentences. In the elaboration of that report devoted entirely to mathematics (College Board 1985), only an additional half page is given to geometry. This is particularly noteworthy compared to the detail and careful consideration given geometry in an earlier report of a committee of the same organization (CEEB 1959).

A recent CBMS report contains a feature that might be adopted in other committee reports; the recommendations, though brief, are connected. The result conveys a flavor not found in the reports cited above.

> We recommend that classes work through short sequences of rigorously-developed material, playing down column proofs, which mathematicians do not use. These proof sequences should be preceded by some study of logic itself. Important theorems not proved can still be explained and given plausibility arguments, and problems involving them can be assigned. The time which becomes available because proofs are de-emphasized can be devoted to study of algebraic methods in geometry, analytic geometry and vector algebra, especially in three dimensions. Work in three dimensions is essential if one is to develop any pictorial sense of relations between many variables, and handling many variables is essential if one is to model phenomena realistically.
>
> There is much room for using computers in geometry. The power of graphics packages makes it much easier for students to get a visual sense of geometric

concepts and transformations. The need to use algebraic descriptions of geometric objects when writing graphics programs reinforces analytic geometry. Finally, the algorithmic thinking needed to write programs bears much resemblance to the thinking required to devise proofs. (CBMS 1983, p. 4)

The Curriculum Dilemma

Allendoerfer noted the fundamental dilemma that underlies the curriculum problem:

> In geometry . . . there is not even agreement as to what the subject is about. Oswald Veblen said, quite seriously, that "geometry is what geometers do." (1969, p. 165)

Thus we see a major reason for the curriculum problem. When specialists cannot come to agreement on the very nature of their subject, it is difficult for government agencies, schools, and teachers to make decisions regarding the geometry to be taught to, and learned by, students. For better or worse, the status quo remains.

Three further suggestions are offered here for resolving the curriculum dilemma:

5. *Clarify the semantics used in discussions of geometry.* Here are three examples that typify the confusion.

a) What is *Euclidean* geometry? In the late sixties, Art Coxford and I wrote a secondary school text entitled *Geometry—a Transformation Approach* (1971). By *transformation approach,* we mean that general definitions of congruence and similarity are given in terms of transformations and that symmetry and other transformation-related ideas are overtly used in the deductive system. As an identifier, *transformation approach* served to distinguish our work. It even may have helped to reinforce the realization that there are many choices available to approaching geometry. However, in place after place, our approach was contrasted with Euclidean geometry, as if it were non-Euclidean.

The contrast of Euclidean geometry with transformation geometry or coordinate geometry reflects more than a semantic difficulty. It reflects confusion between the method of proof and what is proved, between the approach and the results. To virtually all geometers and other mathematicians, Euclidean geometry is a mathematical system that yields the theorems that were put forth by Euclid in his *Elements*. A student is studying Euclidean geometry when he or she is studying the definitions, postulates, or theorems of the geometry. One can take a coordinate approach to Euclidean geometry, or a transformation approach, or a vector approach, or other approaches. Virtually all geometry done in schools today is Euclidean geometry. If we are studying the proofs found in the *Elements,* then we are studying *Euclid's* geometry, one way of approaching Euclidean geometry. The approach usually used in today's high school geometry is a combination of the approaches of Euclid, Legendre, Hilbert, and Birkhoff.

The geometry of Lobachevski, and the equivalent geometry of Bolyai, are the classic non-Euclidean geometries, but there are other geometries that are not Euclidean. The geometry of the surface of the earth is Euclidean if a great circle is considered as a great circle, non-Euclidean elliptic geometry if a great circle is considered as a line. Finite geometries are not Euclidean. The major point to understand here is that few schoolbooks today give significant attention to any geometry that is not Euclidean and that virtually all the alternative approaches that have been suggested for school geometry are still approaches to Euclidean geometry.

b) What are *formal* and *informal,* or *intuitive,* geometry? Throughout this century, a common recommendation has been to "do" informal, or intuitive, geometry in elementary and junior high schools and formal geometry in senior high schools. The distinction is thought to be clear, and it *is* clear at the extremes. Discussing alternative proofs of the Pythagorean theorem is formal geometry. A kindergartner who draws a square is engaging in informal geometry.

But consider the following: A child is asked to draw an isosceles triangle. Drawing seems to be informal geometry, but why would one want to identify isosceles triangles except to set up a formal system either then or later? The child is asked for properties of all isosceles triangles. This requires generalization. In algebra, generalizations are usually considered formal. The child notes that the two angles at the base are congruent. The teacher explains that one could fold the isosceles triangle across a symmetry line to verify this. Folding, still considered informal, is very much like Pappus's proof of the base angles theorem, and there exists a proof by means of reflections and symmetry that is quite analogous to folding. Where does informality stop and formality begin? The phrase "informal proof" reflects the confusion.

The distinction often made between formal and informal geometry leads one to think that there are only two levels of discourse and that these levels are distinct. Such oversimplifications make it more difficult to sequence geometry experiences in a productive manner. Furthermore, they carry the inference that the formality or informality of an idea is independent of the learner. This contradicts what we know from other areas of mathematics. Numbers may be formal to the preschool child but become tools of informal work in the later elementary school. Dienes and Golding (1967) demonstrated that geometry usually considered to be very formal could be made concrete. In fact, one goal in teaching a formal idea could be to change the learner's perception of the idea from formal to intuitive.

In the mathematics education community, we have generally recognized that what is a problem to one person may not be a problem to another. It seems that the same should go for the word *formal*. It might be best to avoid the words *formal* and *informal* and their analogues *abstract* and *concrete,* or *intuitive,* except as they might apply to the state of an individual learner in encountering a concept. Such a practice would bring us close to the views of the van Hieles, who have distinguished at least four *levels of thought* in the learning of geometry (see the article by Crowley in this yearbook).

c) What is a *proof* and when is it valid? There are high school geometry texts in use today that give a "plan for proof" and a "proof" whose only difference seems to be that the former is written in a paragraph and the latter in two columns. The notion that the format of a proof determines its validity is a notion peculiar to high school geometry within our mathematics curriculum.

Valid proofs are often associated with the idea of *rigor.* In many classrooms, there is a de facto definition: a proof is rigorous if there is a reason given for each step. Yet, among mathematicians, rigor varies depending on time and circumstance, and few proofs in mathematics journals meet the criteria used by secondary school geometry teachers. Generally one increases the rigor only when the result does not seem to be correct. (A fascinating discussion in this regard is given by Grabiner [1974] and by Davis and Hersh [1981, chap. 2].)

Many teachers think certain concepts and terminology associated with proof are commonly accepted or understood. Yet some basic questions are not easily answered. Can one have a proof if there are no postulates explicitly stated? Can there be a valid proof if one is not working within a mathematical system? Is a *theorem* a statement that (*a*) has been proved, (*b*) is about to be proved, or (*c*) can be proved, or does the meaning of *theorem* change depending on the content and the needs of the teacher? In practice, we have not come to agreement regarding answers to these questions, and discussion of them would seem to be a prerequisite for resolving the continuing conceptual dilemmas in school geometry.

6. *Refine the level, quality, and quantity of discourse in discussions of the geometry curriculum.* For example, a step toward resolving the dilemma of alternative approaches might be to change the level of discourse from geometry in general to specific topics. We might follow the lead of two recent publications. An entire chapter of *Computing and Mathematics* (Fey 1984) is devoted to the relationships between geometry study and computers. A most ambitious effort is to be found in *Didactical Phenomenology of Mathematical Structures* (Freudenthal 1983), in which almost half the chapters are devoted to exploring the relationships between the geometry of the real world, the mathematician's geometry, and the geometry students encounter in school.

Understanding these readings requires a background in geometry beyond

that found in school geometry. The background of many teachers is insufficient to understand the nature of the curriculum problem. Not only are
many teachers unprepared to teach the content, they are also unprepared
to realize that there might be viable choices. *We will not be able to work
from problems to solutions in school geometry without knowledgeable teachers.*

7. *Analyze, from a curricular perspective, the various ways of conceptualizing geometry.* We have already noted that even geometers do not agree on
the nature of their subject. As educators, it is not our job to resolve such
issues. Instead, the curriculum should reflect the different ways in which
people view geometry. These different ways tend to lead to different goals
and rationales that have been given for studying geometry and lead to
different criteria for the understanding of geometry. I term these ways of
viewing geometry *dimensions,* for reasons that are explained below.

Dimension 1: Geometry as the study of the visualization, drawing, and
            construction of figures

In the early grades, we ask children to draw circles or rectangles or
parallel lines. Later we may ask for images under various transformations:
reflections, rotations, size changes. Students who take a course in "mechanical drawing" learn a great deal of geometry, much of it not found in the
mathematics curriculum. We may allow any equipment or restrict the equipment to straightedge and compass. The impossibility of the trisection of an
angle with this restricted equipment is one of the more difficult ideas to
comprehend in this dimension.

Some psychologists believe aspects of this dimension of geometry are not
affected by experience. Intelligence tests often have spatial perception
items: Given a sequence of visual patterns, find the next. Tell what a figure
looks like after being turned. Count numbers of cubes on which visible
cubes lie. Since virtually every other aspect of the intelligence of most
people is affected by instruction or study, it is doubtful that these aspects
are any different, and what we teach would likely affect students' abilities
to visualize. But visualization and drawing are generally neglected in the
study of geometry.

Dimension 2: Geometry as the study of the real, physical world

When we see the regularity of hexagons in a natural beehive, we ask: Do
bees know geometry? An artist may use geometrical insights overtly in
painting or sculpture. A carpenter may have little geometry training but can
build a house using measuring and rules of thumb. At the other end of the
spectrum, astrophysicists use sophisticated geometry in models of the structure of the universe.

An elementary book exemplifying this view of geometry is Stevens's *Patterns in Nature* (1974), in which geometric ideas such as spirals and branching
are seen in a variety of physical situations. At a higher mathematical level,

Newton's genius was to realize that the physical world behaved in ways that could be expressed by mathematical formulas. Recent contributions to mathematical theory in this area are the landmark books of Mandelbrot (1977, 1983).

Even though geometry evolved from the physical world, connections with that world are relatively ignored in all but the most elementary schoolbooks. Even when found in those books, the real-world connections to geometry tend not to have much direction. Sequencing these connections is an unsolved curricular problem.

Dimension 3: Geometry as a vehicle for representing mathematical or other concepts whose origin is not visual or physical

Geometric representations of nongeometric ideas are familiar to everyone. The number line pictures real numbers. Circle and other graphs display numerical information. The genius of Descartes and Fermat was to picture ordered pairs of numbers as points and to picture solutions to equations involving two variables as lines and curves. We take advantage of coordinate graphing throughout secondary mathematics. Even the concept of symmetry, with its origin in the real world, is often first introduced to students in algebra as a property of the graphs of certain functions or relations. In calculus, area is invaluable as an illustration of what is meant by the evaluation of an integral.

Sometimes the geometry of physical objects is used to understand mathematical concepts. Geoboards can be employed to represent geometrical figures or the coordinate plane, Dienes blocks picture numeration, and Cuisenaire rods help to visualize addition and subtraction.

These examples illustrate the power of geometry to represent ideas from arithmetic, algebra, and analysis. Computer graphics enhance that power. Yet later high school mathematics and computer science courses tend to ignore the geometry students have had before. No doubt we will have to change what we teach in geometry in order for it to be more amenable in this dimension, whose importance is steadily growing.

Dimension 4: Geometry as an example of a mathematical system

Historically, geometry was the first branch of mathematics to be organized logically. In fact, until recent centuries, it was the only branch of mathematics to be so organized. This history affects the geometry curriculum: of all mathematical areas, justifying, discussing logic and deduction, and writing proofs are major goals only in geometry.

Ideas of logic and deduction need not wait until secondary school. Even preschool children understand some aspects of indirect proof. A child can determine that if a ball is behind *A, B,* or *C* and is not behind *A* or *C,* then it must be behind *B.* Direct proof is more difficult, but children can still make logical conclusions. Although it is true that the child's notion of logic is very much incomplete, it is still important to give children experience in drawing inferences. Part of the difficulty in dealing with a systematic approach to geometry in secondary school is surely due to the ignoring of any sort of system earlier.

Though we normally think of this dimension as a senior high school concern, it shapes the entire geometry experience of students. In grades K–8, we name alternative interior angles, or trapezoids, or hypotenuses only because later these terms appear in theorems. Throughout school mathematics, theorems that cannot be proved by standard methods of Euclidean geometry are omitted. For example, the isoperimetric inequality (of all plane figures with a given perimeter, the circle has the most area) is almost always ignored. Yet this theorem is important in the real world and could help students understand the differences between perimeter and area.

When a theorem comes early in the deductive system, then it is studied, even if it is unimportant in the real world or in later mathematics. Why else spend much time on the base angles of an isosceles triangle? Inversely, material that does not come early in the proof sequence is delayed. High school geometry students often study area, volume, three-dimensional geometry, transformations, and coordinates only after they have studied proof, and as a result, most of these ideas do not get the attention they deserve.

The wide extent to which the mathematical-underpinnings dimension dominates geometry study is a weakness of our curriculum. A greater balance is needed.

In summary, there are four major dimensions of geometry: geometry as the visualization, construction, and measurement of figures (the *measurement-visualization dimension*); geometry as the study of the real, physical world (the *physical real-world dimension*); geometry as a vehicle for representing other mathematical concepts (the *representation dimension*); and geometry as an example of a mathematical system (the *mathematical-underpinnings dimension*). Two other dimensions of geometry are occasionally represented in school mathematics but are less important than the four discussed here. A sociocultural dimension deals with the history and development of ideas. A cognitive dimension of understanding, involving one's mental images and cognition, is particularly studied by psychologists.

From a curricular perspective, these different ways of viewing geometry suggest *dimensions of understanding* because the learning of each dimension is relatively independent of the learning of the others and because each dimension contains some ideas that are easy to assimilate and other ideas that are difficult. For this reason the dimensions cannot be rigidly ordered in the curriculum. Like geometric figures themselves, many concepts are multidimensional.

One multidimensional area (no pun intended) is measurement. We need measurement to draw figures accurately, to study the real world, to represent figures, and to understand how geometric properties are related. The per-

vasiveness of measurement in all these dimensions of understanding sub-
stantiates its importance in the geometry curriculum.

An education in geometry that ignores any of these dimensions is too
narrow to be tolerated. Geometry requires the drawing of simple figures
and the interpretation of visual patterns. These patterns continually interact
with the physical world, with other parts of mathematics, and may be logically interrelated with each other in various ways.

Many of the current complaints about geometry can be interpreted as the
natural result of our having taken a one-dimensional view (the mathematical-underpinnings dimension) toward the understanding of geometry concepts (Hoffer 1981). The standard approaches of today may be appropriate
for an understanding of the mathematical system aspect of geometry but
work poorly for understanding its relationship to the real world and its
representation of other mathematics. Computers have changed drastically
the potential drawings and constructions one can make. (For example, the
manuscript for this article was typed on a computer on which a noncircular
ellipse is easier to draw than a circle.) Approaches that integrate geometry
with other mathematics tend to work well for understanding geometric representations of other mathematics but lose sight of geometry as exemplifying
a mathematical system.

Furthermore, these dimensions apply to all levels of study. Logical interrelationships should not be kept hidden from the elementary and junior high
school student. The making of drawings and applications to the real world
should not stop at the geometry teacher's door or be covered only with
slower students. Geometric representations should be a focus of concern
from arithmetic through calculus.

Summary

Geometry is faced with performance and curriculum problems. There is
no geometry curriculum at the elementary school level. As a result, students
enter high school not knowing enough geometry to succeed. There is a
geometry curriculum at the secondary level, but only about half of the
students encounter it, and only about a third of these students understand
it. After the one-year geometry course, geometry is relatively ignored at
both high school and college levels. Consequently the geometry experiences
of both students and teachers are inadequate.

Solutions to these well-known and long-identified problems are confounded by fundamental dilemmas. To improve performance requires more
geometry study, which requires increased numbers of more knowledgeable
teachers, which requires that more people want to study geometry, a desire
usually associated with better performance. To improve the curriculum requires that decisions be made on the inclusion or exclusion of geometry

topics and concepts. However, there is no general agreement regarding the global goals of geometry study in schools, and there has been little careful discussion of the reasons for teaching specific topics.

A number of suggestions are offered for resolving these dilemmas:

1. Specify an elementary school geometry curriculum by grade level.
2. Do not keep students from studying geometry merely because they are poor at arithmetic or algebra.
3. Require a significant amount of competence in geometry from all students.
4. Require that all prospective teachers of mathematics, elementary or secondary, study geometry at the college level.
5. Clarify the semantics used in discussions of geometry; avoid such words as *approach* or *informal* as if they were well defined.
6. Refine the level, quality, and quantity of discourse in discussions of the geometry curriculum.
7. Analyze, from a curricular perspective, the various ways of conceptualizing geometry.

A way of conceptualizing geometry as involving four dimensions of understanding has been put forth, and an outline of an analysis of the concept of congruence has been offered from this perspective.

It is important that we take steps to resolve the continuing dilemmas in school geometry. Geometry is too important in the real world and in mathematics to be a frill at the elementary school level or a province of only half of all secondary school students.

REFERENCES

Allendoerfer, Carl B. "The Dilemma in Geometry." *Mathematics Teacher* 62 (March 1969): 165–69.

Carpenter, Thomas P., Mary M. Lindquist, Westina Matthews, and Edward A. Silver. "Results of the Third NAEP Mathematics Assessment: Secondary School." *Mathematics Teacher* 76 (December 1983): 652–59.

College Board. *Academic Preparation for College: What Students Need to Know and Be Able to Do.* New York: College Board, 1983.

———. *Academic Preparation in Mathematics: Teaching for Transition from High School to College.* New York: College Board, 1985.

College Entrance Examination Board, Commission on Mathematics. *Program for College Preparatory Mathematics.* New York: CEEB, 1959.

Conference Board of the Mathematical Sciences. *The Mathematical Sciences Curriculum K–12: What Is Still Fundamental and What Is Not.* Report to the NSB Commission on Precollege Education in Mathematics, Science, and Technology. Washington, D.C.: CBMS, 1983.

Coxford, Arthur F., and Zalman Usiskin. *Geometry: A Transformation Approach.* River Forest, Ill.: Laidlaw Brothers, 1971.

Davis, Philip J., and Reuben Hersh. *The Mathematical Experience*. Boston: Birkhaüser Boston, 1981.

Dienes, Zoltan P., and E. W. Golding. *Geometry through Transformations*. Vols. 1–3. New York: Herder & Herder, 1967.

Fehr, Howard F., Frank M. Eccles, and Bruce E. Meserve. "The Forum: What Should Become of the High School Geometry Course?" *Mathematics Teacher* 65 (February 1972): 102–3, 151–54, 165–69, 176–81.

Fey, James T., ed. *Computing and Mathematics: The Impact on Secondary School Curricula*. Reston, Va.: National Council of Teachers of Mathematics, 1984.

Freudenthal, Hans. *Didactical Phenomenology of Mathematical Structures*. Hingham, Mass.: D. Reidel, 1983.

Grabiner, Judith. "Is Mathematical Proof Time-Dependent?" *American Mathematical Monthly* 81 (April 1974): 354–65.

Grünbaum, Branko. "Shouldn't We Teach GEOMETRY?" *Two-Year College Mathematics Journal* 12 (September 1981): 232–38.

Hoffer, Alan. "Geometry Is More Than Proof." *Mathematics Teacher* 74 (January 1981): 11–18.

Mandelbrot, Benoit B. *The Fractal Geometry of Nature*. San Francisco: W. H. Freeman & Co., Publishers, 1983.

————. *Fractals: Form, Chance, and Dimension*. San Francisco: W. H. Freeman & Co., 1977.

National Academy of Sciences and National Academy of Engineering. *Science and Mathematics in the Schools: Report of a Convocation*. Washington, D.C.: NAS-NAE, 1982.

National Advisory Committee on Mathematical Education (NACOME). *Overview and Analysis of School Mathematics: Grades K–12*. Washington, D.C.: Conference Board of the Mathematical Sciences, 1975. Available from NCTM.

National Assessment of Educational Progress. *The Third National Mathematics Assessment: Results, Trends and Issues*. Report No. 13-MA-01. Denver: Educational Commission of the States, 1983.

National Council of Supervisors of Mathematics. "Position Paper on Basic Skills." [Distributed to members January 1977.] *Mathematics Teacher* 71 (February 1978): 147–52.

National Council of Teachers of Mathematics. *An Agenda for Action*. Reston, Va.: NCTM, 1980.

————. *Geometry in the Mathematics Curriculum*. Thirty-sixth Yearbook. Reston, Va.: NCTM, 1973.

National Institute of Education. *The NIE (Euclid, Ohio) Conference on Basic Mathematical Skills and Learning*. Vols. 1 and 2. Washington, D.C.: NIE, 1975.

Organization for Economic Cooperation and Development. *New Thinking in School Mathematics*. Report of the Royaumont seminar, edited by Howard Fehr. Paris: OECD, 1960.

Romberg, Thomas. *School Mathematics: Options for the 1990s*. Chairman's report of a conference. Washington, D.C.: Office of Educational Research and Improvement, U.S. Department of Education, 1984.

Senk, Sharon L. "How Well Do Students Write Geometry Proofs?" *Mathematics Teacher* 78 (September 1985): 448–56.

Senk, Sharon, and Zalman Usiskin. "Geometry Proof-Writing: A New View of Sex Differences in Mathematics Ability." *American Journal of Education* 91 (February 1983): 187–201.

Stevens, Peter S. *Patterns in Nature*. Boston: Little, Brown, 1974.

Usiskin, Zalman. *Van Hiele Levels and Achievement in Secondary School Geometry*. Chicago: Department of Education, University of Chicago, 1982.

3

Implications of Computer Graphics Applications For Teaching Geometry

James R. Smart

TEACHING mathematics will never be the same again because of computer science; teaching geometry will never the same again because of computer graphics. The geometry teacher who learns computer graphics will realize the newer field is actually a major new application of geometry. At the same time, the teacher's attitude about geometry will also be changed because of the way geometry is used in computer graphics. Geometry has now become an especially practical subject in the mathematics curriculum because many geometric concepts are essential for computer graphics.

Implications for Teaching Geometry

The following implications suggest nine ways in which new applications of computer graphics can lead to changes in the way geometry is taught. These observations necessarily involve opinions along with facts, so it is hoped that the reader will take these ideas as the basis for discussion as well as action.

1. Geometry students need to realize that the intuitive idea of a point as a very small dot has a particular significance in terms of computer equipment and the concept of a pixel. Recall that a pixel (picture element) indicates the smallest area that can be represented on the computer screen. The part of the screen lighted to represent a point in geometry actually is a very small region. The graphics screen consists of a grid of pixels, often referred to as dots or points. A picture is created by selecting and lighting (energizing) pixels to form the desired pattern. Because of this fact, students need to understand that applications involving points in computer graphics normally involve only points having integral coordinates. Also, students need experience with figures outlined in discrete points as well as those drawn with solid lines.

2. The technical vocabulary of geometry, such as the distinction between *line* and *segment,* or the use of the word *projection,* may be used in a looser sense in applications than within mathematics. Mathematics teachers and students must be flexible enough to understand and accept this fact.

3. Mathematics students in both geometry and algebra must realize that showing the *x*- and *y*-axes with the same scale on both axes, and with positive directions to the right and upward, is a convention that is not necessary. Some computer screens have the *y*-axis pointing downward, since a computer prints from top to bottom. Most have different scales for horizontal and vertical axes. Students should become familiar with different methods of showing axes, and teachers should anticipate that systems of axes in different geometries are not always like those in elementary algebra. For example, the coordinate axes in projective geometry do not meet at right angles, nor is there a uniform scale along an axis.

4. The way slope is handled in computer graphics is a good illustration of the idea that mathematics teachers should be careful to avoid teaching things as if they were always true when they are true only part of the time. Students should not learn that a line with a slope of 1 always goes upward to the right at a 45-degree angle. The way it looks depends on the orientation and scale of the axes. For example, if the *y*-axis on a particular computer screen is positive downward and if the difference in the way the hardware creates horizontal and vertical spacing results in about 500 pixels in each horizontal line and about 250 pixels vertically, then a line through the origin with a slope of 1 appears as in figure 3.1.

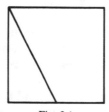

Fig. 3.1

5. Geometry students should expect some "distortion" of geometric figures on a computer screen. For example, the graph of a circle may look elliptical, or the graph of a curve reflected about a line may not look symmetric. The implication is that students should be taught in such a way that they can accept several representations as all being correct. Even though it is possible to modify computer programs so that the resulting figures look the same as they do in geometry books (with little or no "distortion"), this repeated modification is not good educational practice and causes students to be less open-minded about accepting different possibilities.

6. Because they are extremely useful in relatively elementary applications of geometry, particularly in computer graphics, concepts of transformations and their representation as matrices probably should be introduced to the average mathematics student at an earlier grade level than is now done. The brief example that follows shows how several ideas lead to the use of a matrix equation.

In computer graphics, a point can be represented by a matrix. For example, the matrix [3   5] indicates the point (3, 5). The data base of points can be manipulated by means of matrix multiplication, with a 2 × 2 matrix used to represent a transformation in two dimensions. For example, the matrix $\begin{bmatrix} a & 0 \\ 0 & 1 \end{bmatrix}$ represents a scale change in the x direction, and the matrix $\begin{bmatrix} 0 & 1 \\ 1 & 0 \end{bmatrix}$ represents a reflection about the line $y = x$. The result of reflecting a polygon with vertices (3,6), (4,3), (6,2), and (7,5) about the y-axis can be found by using the following matrix equation:

$$\begin{bmatrix} 3 & 6 \\ 4 & 3 \\ 6 & 2 \\ 7 & 5 \end{bmatrix} \begin{bmatrix} -1 & 0 \\ 0 & 1 \end{bmatrix} = \begin{bmatrix} -3 & 6 \\ -4 & 3 \\ -6 & 2 \\ -7 & 5 \end{bmatrix}$$

7. Because of their particular importance in the applied geometry of computer graphics, homogeneous coordinates should be studied at a lower level in the mathematics sequence than the university-level modern geometry course that many students never reach.

Recall that the Euclidean point with coordinates (x, y) can be represented in homogeneous coordinates as (x_1, x_2, x_3), where $x = x_1/x_3$ and $y = x_2/x_3$. One use of homogeneous coordinates in computer graphics is to make possible the representation of large numbers in a microcomputer program, since values are given as ratios. A second use is to extend the types of transformations that can be represented by square matrices. For example, a translation cannot be represented as a 2 × 2 matrix, but it can be represented as a 3 × 3 matrix using homogeneous coordinates. As a specific illustration, a translation of four units in the positive x direction is represented thus:

$$\begin{bmatrix} 1 & 0 & 0 \\ 0 & 1 & 0 \\ 4 & 0 & 1 \end{bmatrix}$$

8. The early need for the visualization of three-dimensional drawings on a computer screen, coupled with teacher experiences that seem to indicate student weaknesses in this skill, suggests strengthening the strand of three-

dimensional geometry throughout the curriculum. It does not seem sufficient to include a brief amount of work in three dimensions at the end of courses in plane geometry. Applications of computer graphics involving three-dimensional objects are very common. They require the use of transformations represented by 4 × 4 matrices when homogeneous coordinates are used. For example, the equation

$$[2 \quad 1 \quad -3 \quad 1] \begin{bmatrix} 1 & 0 & 0 & 0 \\ 0 & 1 & 0 & 0 \\ 0 & 0 & 1 & 0 \\ 2 & 3 & 4 & 1 \end{bmatrix} = [4 \quad 4 \quad 1 \quad 1]$$

shows a translation in three dimensions. Even more significant applications use more general transformations to create perspective drawings. For example, a matrix of the form

$$\begin{bmatrix} 1 & 0 & 0 & a \\ 0 & 1 & 0 & b \\ 0 & 0 & 1 & c \\ 0 & 0 & 0 & 1 \end{bmatrix}$$

can be used to change an ordinary picture of a cube, with opposite edges drawn parallel, to a three-point perspective drawing as shown in figure 3.2, with opposite edges meeting.

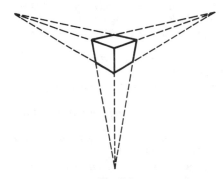

Fig. 3.2

9. Because of their prominence in computer graphics and many other areas, parametric and vector forms for curves and surfaces should be featured in the study of analytic geometry. These are the forms customarily employed when people wish to use mathematical equations in applications. For example, the vector form for a helix is

$$\mathbf{P(t)} = a \cos t \, \mathbf{i} + a \sin t \, \mathbf{j} + bt \, \mathbf{k}.$$

Conclusion

Whenever possible, the many applications of computer graphics that have been described here should be inserted in regular geometry and analytic geometry courses and texts, so that students can see geometry as a very practical branch of mathematics that is essential to computer graphics.

Although sophisticated hardware and software can make it possible to carry out many projects in computer graphics without much background in mathematics or without much apparent use of mathematics, no one should be misled into thinking that the study of mathematics is therefore unimportant. As mathematics teachers, we should be interested in educating the creative people who produce the hardware and software and in producing people who understand the mathematics underlying the applications.

FOR FURTHER READING

Abelson, Harold. *Apple Logo*. Peterborough, N.H.: Byte/McGraw-Hill, 1982.

Artwick, Bruce A. *Applied Concepts in Microcomputer Graphics*. Englewood Cliffs, N.J.: Prentice-Hall, 1984.

Borland International, Inc. *Turbo Pascal Version 3.0 Reference Manual*. Scotts Valley, Calif.: Borland International, 1985.

Demel, John T., and Michael J. Miller. *Introduction to Computer Graphics*. Monterey, Calif.: Brooks/Cole, Engineering Division, 1984.

Dyer, Thomas A., and Margot Critchfield. *A Bit of BASIC*. Reading, Mass.: Addison-Wesley Publishing Co., 1980.

Foley, James D., and Andries Van Dam. *Fundamentals of Interactive Computer Graphics*. Reading, Mass.: Addison-Wesley Publishing Co., 1982.

Glassner, Andrew S. *Computer Graphics User's Guide*. Indianapolis: Howard W. Sams & Co., 1984.

Harper, Dennis O., and James H. Stewart. *RUN: Computer Education*. Monterey, Calif.: Brooks/Cole Publishing Co., 1983.

Harrington, Steven. *Computer Graphics, a Programming Approach*. New York: McGraw-Hill, 1983.

Hartman, Roy A., and Louis Vrooman. *Computer Graphics*. College Station, Tex.: Creative Publishing, 1983.

National Council of Teachers of Mathematics. *Computers in Mathematics Education*. 1984 Yearbook. Reston, Va.: The Council, 1984.

Newman, William M., and Robert F. Sproull. *Principles of Interactive Computer Graphics*. New York: McGraw-Hill, 1979.

Rogers, David F., and J. Alan Adams. *Mathematical Elements for Computer Graphics*. New York: McGraw-Hill, 1976.

Waite, Michael. *Computer Graphics Primer*. Indianapolis: Howard W. Sams & Co., 1979.

4

Can Geometry Survive in the Secondary Curriculum?

Ivan Niven

THE advent of computer science and its mathematical handmaidens, algorithms and discrete mathematics, has brought to center stage the lively debate on what should be included in the secondary curriculum. As a consequence, advocates of geometry are under great pressure to justify its place in the curriculum. My purpose here is to discuss ways to make geometry a more attractive subject at the secondary level, so that it becomes easier to defend its rightful place in the curriculum.

Certainly there can be no doubt about the importance of geometry in its supporting role not only in other parts of mathematics but also in such areas as engineering, architecture, physics, and astronomy. There is no need here to press the point about the significance of the subject. Instead, let us turn to the matter of enhancing the attractiveness of the first course in geometry; to this end I offer a list of nine proposed recommendations. In some of these, the reader will notice that I am simply urging the extension and strengthening of trends that have already begun.

Recommendation 1. Teach beginning geometry in the same way beginning algebra and beginning calculus are taught, without excessive emphasis on rigor.

This is the key point, so it will be elaborated on at some length. It is traditional, although fortunately the approach has declined slightly, to use geometry as a vehicle to reveal to students the nature of mathematics as an axiomatic system. That is, geometry is presented as a system on which an edifice of theorems is built on certain undefined terms and postulates. The consequence for many students is that they consider axioms, theorems, and proofs a part of geometry only, not a part of subjects like algebra, trigonometry, and calculus.

Do postulates and theorems exist in algebra? They certainly do, but for some reason they are called by different names. In the real number system it is a basic assumption that if $ab = 0$, then $a = 0$ or $b = 0$ or both. This postulate is usually called by some informal name like a "rule" and is displayed in a rectangular box in italics or block type for emphasis. Similarly, one of the basic theorems in algebra is this: Given constants a, b, c, $a \neq 0$, the equation $ax^2 + bx + c = 0$ has two solutions that can be readily exhibited. But that result is just called the quadratic formula. A theorem? Yes, it really is a theorem, but nobody calls it that. The subject is developed in a less formal way, fortunately.

Similarly, the axioms and theorems in calculus are usually called by other names. For example, it is a basic assumption needed repeatedly in the theory of limits that a bounded increasing sequence such as 1/2, 2/3, 3/4, 4/5, . . . has a limit, namely 1. But this postulate is given in many calculus books simply as an important principle to be followed. As to theorems, some calculus books list very few theorems other than the fundamental theorem of calculus and the mean value theorem. The other theorems are there, but they are presented more casually.

The famous geometer H. S. M. Coxeter commented in an interview (Logothetti 1980) that in his opinion the teaching of geometry has been more effective in Europe than in Canada or the United States. Asked why, Coxeter replied:

> I think because there was a tradition (in Canada and the United States) of dull teaching; perhaps too much emphasis on axiomatics went on for a long time. People thought that the only thing to do in geometry was to build a system of axioms and see how you would go from there. So children got bogged down in this formal stuff and didn't get a lively feel for this subject. That did a lot of harm. And you see if you have a subject badly taught, then the next generation will have the same thing, and so on in perpetuity. (P. 15)

Coxeter's view is shared not only by many experts in geometry but by mathematicians generally who rebelled against an excess of attention to axiomatic structures. The distinguished applied mathematicians Richard Courant and Herbert E. Robbins (1947) put the matter this way in their great classic:

> The axiomatic approach to a mathematical subject is the natural way to unravel the network of interconnections between the various facts and to exhibit the essential logical skeleton of the structure. . . . But a significant discovery or an illuminating insight is rarely obtained by an exclusively axiomatic procedure. Constructive thinking, guided by the intuition, is the true source of mathematical dynamics. Although the axiomatic form is an ideal, it is a dangerous fallacy to believe that axiomatics constitutes *the* essence of mathematics. (P. 216)

Postulates and axioms have proliferated in geometry books like rabbits in the last two or three decades; authors seem to vie with one another to find or even create new ones. Pasch's axiom has suddenly appeared: *If a line intersects one side of a triangle at any point other than a vertex, it will also intersect another side of the triangle.* Are we not in danger that the students will see geometry as just so much nit-picking? The "square roots postulate" has appeared: *If $a = b \geq 0$, then $\sqrt{a} = \sqrt{b}$*. After observing that the books featuring this postulate seemed to have no need for a corresponding cube roots postulate (if $a = b$, then $\sqrt[3]{a} = \sqrt[3]{b}$), I asked five professors, all specialists in algebra or geometry, if they knew something called "the square roots postulate." They all said no, but one guessed that it must be the statement that if a is positive, then $\sqrt{a}$ exists. When I told them what it was, three said that they preferred the more general substitution postulate: *If $a = b$, then b can be substituted for a, or a for b in any formula or equation*.

Postulates and axioms should not get so much attention in beginning courses. It would suffice simply to say:

> We take the following basic ideas for granted without proof and then build the structure of geometry on them. In some more advanced courses these basic concepts are examined more carefully.

Then a short narrative could be given that two distinct points determine one and only one line, that two lines in a plane either intersect in a point or are parallel, that a line segment has a unique midpoint that divides the segment into two parts of equal length, and so on. All this should be illustrated by diagrams—lots of diagrams.

It has long been recognized that the level of rigor should get more exacting as we proceed from introductory courses to more advanced discussions. Silvanus P. Thompson (1910) wrote in his calculus book:

> You don't teach the rules of syntax to children until they have already become fluent in the use of speech. It would be equally absurd to require general rigid demonstrations to be expounded to beginners in the calculus. (P. 236)

The same principle applies to beginning geometry.

Our main point, then, is to free the teaching of elementary geometry from its traditional role of serving as a general introduction to the axiomatic structure of mathematics. Why should the first course in geometry carry the special burden of illustrating and exemplifying the foundations of mathematics? The van Hiele studies (Mayberry 1983) demonstrate clearly that most of the students are not yet ready for such abstract topics. Let's teach geometry as geometry, in the way that algebra and calculus are taught.

Recommendation 2. Get to the heart of geometry as soon as possible.

One of the eight geometry textbooks on my desk characterizes the theorem of Pythagoras as "perhaps the most famous theorem in all of mathematics." How true! But the famous theorem does not appear until page 338 of that text, and in similar locations in the other seven. Since it is possible to get to that theorem in the first hundred pages, why is it delayed until page 338? It is the foundation stone of so much further work in mathematics itself and in the applications of mathematics; thus we should drive to it as fast as we can.

How is it possible to get to the Pythagorean theorem in the first hundred pages? There is a well-known proof, illustrated in figure 4.1, that does the job. The total area of the square is $(a + b)^2$, but it is also $c^2 + 4(1/2\ ab)$ by adding up the pieces. We get $c^2 = a^2 + b^2$ by equating these two expressions for the total area. To complete this proof, we must verify that the inner figure is a square and not a rhombus. It is easy to check that each angle of the inner figure is a right angle.

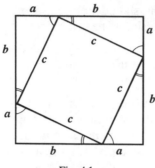

Fig. 4.1

Textbook writers should organize their books so as to reach as quickly as possible such central topics as the Pythagorean theorem, congruence, and similarity.

Recommendation 3. Use the techniques of algebra and analytic geometry as well as the classical Euclidean methods.

If objection is raised that geometry is geometry, not algebra, the rebuttal is that watertight compartments in mathematics should have long since disappeared. The integration of knowledge is a very important matter, and we should not pass up good natural opportunities for unifying different topics. An illustration of the use of simple algebra is given in the proof of the theorem of Pythagoras above, and dozens more of such illustrations could be given.

Analytic geometry provides an introduction to the very important topic of graphing and enables us to prove many results in simpler ways. We give just one example here, an easy proof of the converse of the Pythagorean theorem, using coordinates. We are given a triangle satisfying the relation $c^2 = a^2 + b^2$, where a, b, and c are the lengths of the sides. We are to prove that the angle opposite the length c is a right angle. Taking the vertex opposite the side of length c as the origin $(0,0)$ and the x-axis along the side of length a, we have a second vertex at $(a,0)$. Assign to the third vertex the coordinates (r,s), as shown in figure 4.2.

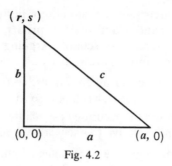

Fig. 4.2

We prove that the vertex (r,s) lies on the y-axis, from which it follows at once that the angle at $(0,0)$ is a right angle. By the distance formula we know that—

$$c^2 = (r - a)^2 + (s - 0)^2 \text{ and } b^2 = (r - 0)^2 + (s - 0)^2,$$

or

$$c^2 = r^2 + s^2 + a^2 - 2ra \text{ and } b^2 = r^2 + s^2.$$

Using $c^2 = a^2 + b^2$, we have

$$r^2 + s^2 + a^2 - 2ra = a^2 + r^2 + s^2$$

and hence $-2ra = 0$ and $ra = 0$. Now a is not equal to zero, since it is the length of the side of a triangle. Hence $r = 0$, and the point (r,s) lies on the y-axis.

Recommendation 4. Use diagrams in all explanations, especially in proofs.

Geometry is a visual subject, so that figures are of fundamental importance. This is not to say, as has been claimed occasionally in recent times, that the course in geometry might well be replaced by computer graphics. Graphs are very instructive, but they are not the whole story—not by a long shot.

Look at figure 4.3, which shows that the sum of the angles of a triangle is 180°. This is accomplished by the use of a line segment drawn through one vertex of a triangle parallel to the opposite side. It is easy to remember such simple, direct arguments.

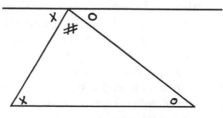

Fig. 4.3

And what is very important, we should not hesitate to use diagrams as a basis for giving explanations and making proofs, certainly at the level of rigor expected in the secondary school. Arguing from a diagram is valid provided two simple conditions are satisfied:

1. The diagrams must cover all situations, so that it might be necessary to have two or more figures, such as an obtuse triangle and an acute triangle.

2. The diagrams must be accurate, especially with respect to the positions of points and lines relative to each other.

Students find it amusing to see examples of phony diagrams being used to "prove" outlandish results. One well-known example is a "proof" that every triangle is isosceles, based on an inaccurate diagram: the intersection point of the bisector of an angle of the triangle and the perpendicular bisector of the side opposite that angle is located far from where it should be, thereby twisting the diagram out of shape. An accurately drawn diagram reveals at once what the trouble is.

Recommendation 5. Relate geometry to the mainstream of mathematics and to the real physical world.

How did the ancient Greeks figure out that the radius of the earth is approximately 3960 miles? By geometry, of course. If you know the angle of elevation of the sun at noon, measured at two different places A and B, where A is a known distance due north of the point B, the radius of the earth can easily be calculated. With some help on the general idea here, students can finish the calculation for themselves.

Here is a second example. The owner of a racing car, wanting to revive his somewhat worn engine, had the cylinders rebored (thereby increasing their diameter). He then reassembled the engine with larger pistons. After the job was complete, it suddenly occurred to him that by enlarging the engine, he might have inadvertently put his car in the next larger racing class. He had been racing in the under 2.5 liters class; if by reboring the cylinders he had enlarged the engine displacement to just over 2.5 liters, the car would be in the under 3.0 liters class. He knew his car would be no competition at all for cars with nearly 3.0 liters of displacement. Almost in a state of panic, he called a mathematics teacher (it happened to be me) and was delighted to learn that his engine still qualified for the smaller class.

The facts needed to answer his question were (1) the number of cylinders, (2) the length of the stroke (unchanged by the alteration), and (3) the diameter of the new pistons, called the bore of the engine. In the standard formula $V = \pi r^2 h$, h is the length of the stroke and r is half the bore.

There is a need for more collections of good applications, especially in textbooks, for easy use by teachers.

Recommendation 6. Eliminate the verbosity, and avoid the deadly elaboration of the obvious.

Consider Proposition 17 from Book V of Euclid's *Elements:*

> If two magnitudes, taken jointly, be proportionals, they shall also be proportionals when taken separately; that is, if two magnitudes taken together have to one of them the same ratio which two others have to one of these, the remaining one of the first two shall have to the other the same ratio which the remaining one of the last two has to the other of these.

What is all this about? With respect to the lengths shown in figure 4.4, it says that

$$\text{if } \frac{a + b}{a} = \frac{c + d}{c}, \text{ then } \frac{b}{a} = \frac{d}{c}.$$

To prove this, simply subtract 1 from both sides of the first equation to get the second. Euclid did not have algebra available, so his proof is rather long, as might be expected.

Fig. 4.4

Has this tradition of wordiness in geometry, this excessive use of words in proportion to the thought, been abandoned since the time of Euclid? Unfortunately, no. Here are two consecutive theorems from a current geometry text:

> If two sides of one triangle are congruent to two sides of another triangle, but the included angle of the first triangle is larger than the included angle of the second, then the third side of the first triangle is longer than the third side of the second.

> If two sides of one triangle are congruent to two sides of another triangle, but the third side of the first triangle is longer than the third side of the second, then the included angle of the first triangle is larger than the included angle of the second.

These statements could have been shortened considerably by direct reference to the sides and angles of specific triangles, as in figure 4.5. Furthermore, the words "and conversely" could have been added to the statement of the first theorem, thereby eliminating the need for the second statement.

Recommendation 7. Postpone or omit the proofs of some difficult theorems.

Many beginning books in calculus quite deliberately and properly step aside from giving full, rigorous proofs of such results as the fundamental theorem of calculus and L'Hôpital's rule. Intuitive or heuristic arguments

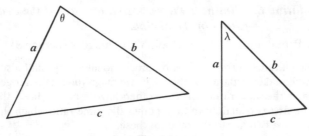

Fig. 4.5

are given that are somewhat persuasive. The same is true for the fundamental theorem of algebra in beginning algebra books.

The same pattern should be followed in the first course in geometry.
Consider, for example, Heron's formula for the area of a triangle with sides
of lengths a, b, c, where s is the semiperimeter, $s = (a + b + c)/2$:

$$[s(s - a)\,(s - b)\,(s - c)]^{1/2}$$

It is difficult to prove this result by elementary geometry but not so difficult
by trigonometry. Ptolemy's theorem is another result that is hard to prove
without advanced methods: If A, B, C, and D are four points in that order
on the circumference of a circle, then $AB \cdot CD + AD \cdot BD = AC \cdot BD$,
where, for example, $AB \cdot CD$ denotes the product of the two lengths AB
and CD.

Of course, we should not abandon all proofs and offer cookbook courses.
One of the glories of geometry has always been that it serves as a model of
careful reasoning, drawing valid conclusions from given information. But if
the proof of a result by elementary methods is excessively long or difficult
and if an easier proof is available by techniques not yet introduced, we
should consider postponing the proof.

**_Recommendation 8. Textbooks should offer lots of problems of intermediate
difficulty for classroom use._**

Current textbooks offer us row after row of extremely easy, almost simple-
minded problems:

A point is given. Is the point contained in just one line, or more than one line?

A few simple problems are needed, of course, to fix the basic definitions
and concepts in the minds of the students. But surely 50 to 100 such problems would suffice for that purpose. Some texts seem to have no problems
of any real difficulty whatsoever.

It is not easy to define "problems of intermediate difficulty," but here are
two:

Let P, Q, R, and S be the midpoints of the sides AB, BC, CD, and DA,
respectively, of a convex quadrilateral $ABCD$.

1. Prove that *PQRS* is a parallelogram.
2. Prove that the area of *PQRS* is half that of *ABCD*.

The question "Do these results hold for a nonconvex quadrilateral *ABCD?*" is not of intermediate difficulty; it is harder.

Recommendation 9. Tell students the full story about trisecting an angle.

This is a special point but of considerable importance because of the widespread misunderstanding about whether or not it is possible to trisect an angle. Is it impossible? Yes, it is, if we impose the severe limitation originated by the ancient Greeks that only compass and *unmarked* straightedge can be used. But if marks are allowed on the straightedge, it is easy to trisect an angle. When the question of the trisection of angles is discussed in a secondary school class, some trisection procedure should be presented to the class, such as the one given below. Otherwise, many students will come away from their geometry course persuaded that it is impossible to trisect an angle. Some of these students may even become "trisection nuts" and "solve" the problem that has confounded mathematicians for centuries!

In figure 4.6 a given angle *ABC* is to be trisected. Select a point *P* on the segment *AB* and draw *PQ* parallel to *BC*, and *PR* perpendicular to *BC*, meeting *BC* at *R*. On the straightedge mark two points, say *S* and *T*, so that the distance *ST* is twice the distance *BP*. (The points *S* and *T* are on the straightedge and so are not shown in fig. 4.6.) Slide the straightedge on the diagram, keeping S on the line segment *PR* and T on the segment *PQ*, and thus locate *H* on *PR* and *K* on *PQ* to satisfy these conditions: *HK* = 2*BP*, and *B*, *H*, and *K* are collinear. The line segment *BK* is a trisector of the angle *ABC*.

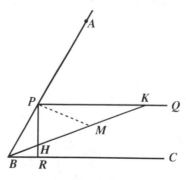

Fig. 4.6

The proof of this, not given here, is at the intermediate level as discussed in the preceding section. The main point needed is that the midpoint *M* of the segment *HK* is equally distant from the vertices of the right triangle *PHK* so that we have two isosceles triangles *BPM* and *PMK*.

Is there something unnatural about putting marks on a straightedge? No, not at all.

In conclusion, let us go back to the question, "Can geometry survive in the secondary curriculum?" The answer is an enthusiastic yes. Geometry can survive in the curriculum if it is approached not too pedantically, with the emphasis on geometry as such and not on axiomatics and foundations.

BIBLIOGRAPHY

Clemens, Stanley R., Phares G. O'Daffer, and Thomas J. Cooney. *Geometry.* Menlo Park, Calif.: Addison-Wesley Publishing Co., 1983.

Courant, Richard, and Herbert R. Robbins. *What Is Mathematics?* 4th ed. New York: Oxford University Press, 1947.

Coxford, Arthur F., and Zalman P. Usiskin. *Geometry, a Transformation Approach.* River Forest, Ill.: Laidlaw Bros., 1971.

Hoffer, Alan R. *Geometry.* Menlo Park, Calif.: Addison-Wesley Publishing Co., 1979.

Jacobs, Harold R. *Geometry.* San Francisco: W. H. Freeman & Co., Publishers, 1974.

Lang, Serge, and Gene Murrow. *Geometry.* New York: Springer-Verlag, 1983.

Logothetti, David. "An Interview with H. S. M. Coxeter, the King of Geometry." *Two-Year College Mathematics Journal* (January 1980): 15.

Mayberry, Joanne. "The van Hiele Levels of Geometric Thought in Undergraduate Preservice Teachers." *Journal for Research in Mathematics Education* 14 (January 1983): 58–69.

Moise, E. E., and Floyd L. Downs, Jr. *Geometry.* Menlo Park, Calif.: Addison-Wesley Publishing Co., 1967.

Jurgensen, Ray C., and Richard G. Brown. *Basic Geometry.* Boston: Houghton-Mifflin Co., 1981.

Jurgensen, Ray C., A. J. Donnelly, and Mary P. Dolciani. *Modern Geometry.* Boston: Houghton Mifflin Co., 1963.

Rhoad, Richard, George Milauskas, and R. Whipple. *Geometry.* Evanston, Ill.: McDougal, Littell & Co., 1981.

Thompson, Silvanus P. *Calculus Made Easy.* 3d ed. London: Macmillan, 1957.

5

Euclid May Stay—
and Even Be Taught

Tommy Dreyfus
Nurit Hadas

EUCLIDEAN geometry has been taught less in recent years than it was twenty years ago. The reason for this decline is to be found not in dissatisfaction with the content but rather in the conceptual difficulties caused by the logical arguments that form the essence of Euclidean geometry. Most of the difficulties that students are observed to have in the classroom relate to the way they organize their thoughts and their construction of logical arguments. In this article, some of these conceptual difficulties will be exposed, strategies for overcoming them will be detailed, and the successful implementation of these strategies in a new geometry course will be reported.

The central idea behind the innovative methods and the new curriculum materials described below is to deal directly and explicitly with the organization of students' thought patterns and their construction of logical arguments. For example, those students who find it hard to understand the necessity of proving statements that appear obvious or well known from earlier classes will receive exercises in which they have to either justify or refute statements of two types: (1) those that appear to be correct but are not, and (2) those that are correct but are not evident. Teachers who have taught Euclidean geometry will recognize many of the difficulties discussed and identify them with similar ones from their own classroom experience.

We are indebted to Alex Friedlander whose contribution to the work described here has been on a par with ours and who should have been our coauthor. (NCTM policy prohibits a person from coauthoring two articles in the same yearbook.) Our work has also been influenced by our colleagues in the Mathematics Group of the Department of Science Teaching at the Weizmann Institute of Science—in particular, Maxim Bruckheimer, Rina Hershkowitz, Emmanuel Kremer, and Naomi Taizi. Some of the writing on this article was done while Tommy Dreyfus was a visiting scholar in the Department of Mathematics at San Diego State University.

Background

The controversy about high school geometry has been going on for years. As a result of the "Euclid must go" syndrome, a wealth of alternative geometry courses has been proposed (see, for example, Freudenthal [1973]). Many eminent mathematicians and experienced mathematics educators have made relevant contributions to the discussion, among them Thom (1973) and Atiyah (1977). We shall not side with one group or the other in this discussion but rather espouse the point of view taken by Fletcher (1970–71), who wrote that

> the problem was not to free Euclid from logical blemish, the problem was to replace it [Euclid's text] by a teaching strategy that was more acceptable. (Pp. 395–96)

Indeed, it has been clearly demonstrated recently that even among senior high school students who receive extensive mathematics training, two-thirds do not understand what a mathematical proof is (Fischbein and Kedem 1982). One attempt at solution was undertaken by Deer (1969), who showed that the prior teaching of an explicit unit on logic had no effect on students' ability to learn to construct proofs independently. It thus appeared that attempts at devising appropriate teaching strategies should be based on a detailed understanding of the cognitive difficulties of students, as evidenced by recent research (Fischbein and Kedem 1982; Vinner and Hershkowitz 1983; Balacheff 1985). In a recent study, Senk (1985) identified student difficulties in writing proofs and recommended research "to identify cognitive and affective prerequisites for doing proofs and techniques for helping students acquire these prerequisites" (p. 455). This article provides a partial answer to Senk's request.

Principles

Our analysis of student difficulties led to the formulation of six principles that may seem obvious to mathematicians, curriculum writers, and teachers but that are not well understood by most students of average ability, as will be shown:

Principle 1. A theorem has no exceptions. A mathematical statement is said to be correct only if it is correct in every conceivable instance.

Principle 2. Even "obvious" statements have to be proved. In particular, a proof may not be built on the apparent features of a figure.

Principle 3. A proof must be general. One or more particular cases cannot prove a general statement. However, one counterexample is sufficient to refute it.

Principle 4. The assumptions of a theorem must be clearly identified and distinguished from the conclusions.

Principle 5. The converse of a correct statement is not necessarily correct.

Principle 6. Complex figures consist of basic components whose identification may be indispensable in a proof.

The majority of these principles are of a logical nature. Basically, they can be implemented in any Euclidean geometry course and for different ability levels. For us, they served as a basis for the pedagogical considerations that guided the development of a ninth-grade course in Euclidean geometry for students of average ability (Friedlander and Hadas 1981). The course has since been adapted for students of higher ability. The six principles have been systematically implemented in the exercises that compose the bulk of the textbook. Each exercise emphasizes the relevant principles throughout in an integrated manner. Most exercises require the use of more than one principle. However, in order to capture the spirit of the course and to elucidate the underlying rationale, we will present sample exercises in six categories corresponding to the principles.

Principle 1. A theorem has no exceptions.

Often, students do not realize that a mathematical statement is said to be correct only if it is correct in each and every conceivable instance in which the assumptions are satisfied. The exercises given below are designed to help clarify this important mathematical principle—that a theorem has no exceptions.

1. Complete to obtain true statements:
   a) A quadrilateral with four equal sides is a _____.
   b) A quadrilateral with three equal sides is a _____.

2. Consider statements (a) and (b) below.
   - If true, explain.
   - If false, find a counterexample.
   a) A quadrilateral with all angles equal is a square.
   b) A quadrilateral with all angles equal is a rectangle.

The misconceptions specifically addressed in these exercises are all too familiar to geometry teachers:

- That a quadrilateral with four equal sides (exercise 1a) must be a square or that a quadrilateral with three equal sides (1*b*) must be a square, a rhombus, or at least an isosceles trapezoid

- That a quadrilateral with "many" equal angles must be regular—that is, it must be a square (exercise 2)

Students should be encouraged to consider different cases before they decide whether a statement is true or false. We found it very helpful for this purpose to let them manipulate small pieces of transparency on which angles, segments, or lines were drawn.

In exercise 1, the use of four congruent segments (fig. 5.1a) or three equal segments and a straight line (fig. 5.1b) may be helpful in erasing some potential misconceptions. Similarly, for exercise 2, assembling four right angles to form a rectangle will help one notice that squares are not the only quadrilaterals with four equal angles (fig. 5.2).

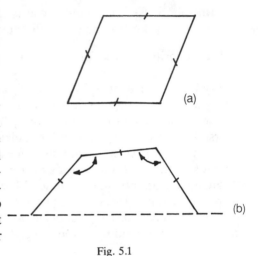

(a)

(b)

Fig. 5.1

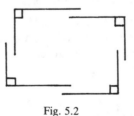

Fig. 5.2

Principle 2. Even "obvious" statements have to be proved.

Visualization and measurement are vital strategies for discovering geometrical properties. In a deductive system, however, each statement has to be based on statements proved earlier, and one cannot prove a statement by relying on figures. For most students, the transfer from informal geometry (where properties are deduced from specific figures or objects) to Euclidean geometry (where formal deduction is required) is a source of considerable confusion. It is therefore necessary to teach and emphasize the newly introduced principle that statements have to be proved. Usually, the need to prove is dogmatically imposed on the student as an abstract principle or as a "rule of the game," and then the student is presented with a host of

immediate, easy-to-visualize properties to prove. Paradoxically, this makes the transition even more difficult, because intuitive visualization confounds the issue. However, we have observed that posing questions and accompanying them with a misleading or inconclusive figure may help the student overcome some of this confusion. It is neither neces-
sary nor helpful to wait for
more advanced topics in geometry to implement this
idea. The following two
problems illustrate this
point:

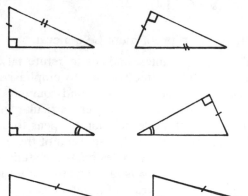

3. Determine whether the triangles in each of the three pairs in figure 5.3 are congruent.

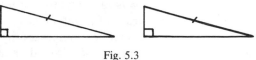

Fig. 5.3

Clearly, visualization is not very helpful and may even be misleading here.

4. In exercises (a), (b), and (c) below, consider this statement: $\triangle AOB \cong \triangle DOC$.

- If true, prove it.
- If false, draw a counterexample.

a) Given $AB = CD$

b) Given $AB \parallel CD$

c) Given $AB \parallel CD$
   $AB = CD$

In a traditional course, students are given an exercise like 4c as a proof problem; that is, the statement is presented as an established fact. As a result, they see little need to prove the statement. The dynamic aspect of

the sequence of problems in exercise 4 may be
further emphasized by letting a freely moving
line create a variety of triangles *ADB* on a
transparency from which the segment *AB* is
lacking (see fig. 5.4).

Principle 3. A proof must be general.

The use of counterexamples to refute false
statements is an effective way to emphasize
principle 3. In their efforts to find counterexamples, students often discover particular instances in which the conclusion happens to be
valid (see exercise 2*a*). A discussion of the examples and counterexamples helps the student
to appreciate that a general deductive proof is
needed to show the truth of a statement.

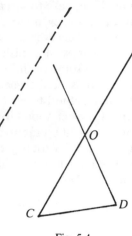

Fig. 5.4

The same principle can be emphasized by posing geometrical "bafflers"
rather than requiring proofs of self-evident and easy-to-visualize facts. The
effect of surprise when a student discovers an unexpected fact can lead her
or him to see the need for a general proof.

5. Consider △*DOG* and △*CAT.*

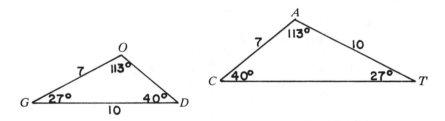

   a) How many pairs of equal sides and angles do the two triangles have?
   b) Are the two triangles congruent? Explain.
   c) Can you find two noncongruent triangles with six pairs of equal sides and
      angles (with no side or angle paired more than once)? Explain.

Here the lack of correspondence between the five given pairs of equal
elements runs against the misconception that three, or at most four, pairs
are sufficient to conclude the congruence of two triangles. The counterexample presented in exercises 5*a* and 5*b* also emphasizes the need for a
general argument in 5*c*.

6. Given: *A* circle with arc *AB*.

   Draw on tracing paper two angles,

   $\alpha = \dfrac{1}{2} \overset{\frown}{AB}$ and $\beta \ (\beta \neq \alpha)$.

   Place one of the angles with its vertex on the given circle so that its sides pass through the given points *A* and *B*. Repeat with the other angle.

   a) In how many different ways can you do this with α?

   b) In how many different ways can you do this with β?

   c) Draw conclusions from your findings.

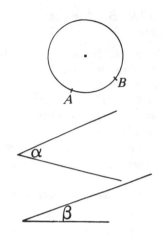

Students are frequently astonished to discover that there are an infinite number of positions for α, whereas for β there are none. The only way to settle the conflict between this discovery and the expected result is by deductive reasoning: α happens to be half of the central angle *AOB*, and any peripheral angle of a different size cannot sustain the arc *AB*.

Principle 4. The assumptions of a theorem must be clearly identified and distinguished from the conclusions.

The distinction between assumptions and conclusions is a prerequisite for proving a theorem. Students must be familiar with the "if-then" structure of a mathematical statement. They need to be able to verify assumptions and conclusions even when they are not clearly separated in the formulation.

7. Put the following sentences in an "if-then" form:

   a) Opposite angles are equal.

   b) A quadrilateral whose diagonals bisect each other is a parallelogram.

8. Given: *KLMN* is a rhombus.

   a) Prove: $\triangle KOL \cong \triangle MOL$

   b) Conclude from (a): $\angle KOL = \angle$ _____

   $\angle KOL =$ _____°

   c) Write a theorem that expresses your conclusions.

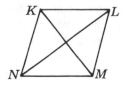

Exercises 7 and 8 present two possible ways of stressing the distinction between assumptions and conclusions: we can ask the student to transcribe "ordinary" sentences into "if-then" statements (exercise 7), or we can reverse the task as in exercise 8.

Principle 5. The converse of a correct statement is not necessarily correct.

Most traditional geometry courses note that the converse of a true statement is not necessarily true. The same courses, however, rarely require the student to recognize and refute any statements, let alone false converses. Therefore, this important logical principle is quickly forgotten, and students may easily get the impression that the converse of a correct statement is always correct.

In our course, we deal quite extensively with false statements in general and with false converse statements in particular.

9. Below we give three correct statements—*a, b,* and *c.*

   • The good news is: You don't have to prove them!
   • The other news is: You have to
         —put them in "if-then" form;
         —find the converse statements;
         —determine whether the converse statements are true or false.

   a) The diagonals of a kite are perpendicular.

   b) A line parallel to a side of a triangle creates a new triangle that is similar to the original one.

   c) An isosceles triangle has two equal angles.

Manipulating tracing paper may help students refute the converse of the first two statements. By moving two perpendicular segments, one can see that the converse of statement 9*a* is false (fig. 5.5a). For the converse of statement 9*b*, manipulating tracing paper may be useful in finding a similar triangle that is created by a line not parallel to any of the sides of the original triangle (see fig. 5.5b).

(a)

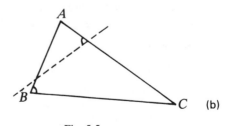

(b)

The converse of statement 9*c* is true, and, accordingly, it has to be proved. This point

Fig. 5.5

was not clear even to W. Emerson, who was, in his time, a well-known mathematician. In the preface to his *Elements of Geometry,* Emerson (1763, pp. vi–vii) presents the following argument:

It is a common practice among geometers, after a propofition is proved, for them to prove the re-verfe of it. But this in many cafes is needlefs and impertinent. . . . For inftance, when the two fides of a triangle are equal, it may be proved, that the two oppofite angles are equal. Or when the two angles of a triangle are equal, it may be proved, that the oppofite fides are equal. But it need not be proved both back and forward.

Principle 6. Complex figures consist of basic components.

Many students tend to link a certain geometrical shape with a specific "standard" position (Vinner and Hershkowitz 1983):

- Isosceles triangles are recognized only when their base is "horizontal" and are generally drawn that way.
- Kites are identified only if their main diagonal is "vertical."
- The corresponding sides of two congruent triangles must be (more or less) parallel for them to be recognized as such.

These and similar misconceptions (or rather, partial conceptions) may be corrected if students are also presented with congruent triangles whose corresponding sides are not parallel (see exercise 1), or with complex figures whose components are not in the standard position. Exercise 10 is an illustration of the latter.

10. Given: *CD* bisects ∠ *ACB*
    AE bisects ∠ *BAC*
    ∠ *DCB* = 36°
    ∠ *BAC* = 72°

    a) Find the size of all the other angles.
    b) Identify all the isosceles triangles in the figure.

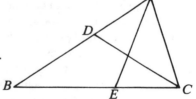

Many students start to rotate the figure to bring the component triangles into an easily recognizable position. This visual strategy, however, is hardly sufficient for finding the complete answer, namely, identifying all eight isosceles triangles!

Implementation and Evaluation

It is worth stressing that the uniqueness of the course described here lies in its approach, not in its content. The curriculum follows the standard sequence for a first course in Euclidean geometry, but strategies, based on the six principles, have been specifically designed to deal with students' difficulties. The development of the course took about four years. During this entire period, classroom teachers were actively involved in the work, and much of the material was rewritten repeatedly on the basis of their reports and classroom observations. In-service workshops and courses were organized to familiarize the teachers with the approach. The reactions to the meetings were excellent: Teachers wrote favorable comments and also reported a considerable increase in student motivation to learn geometry.

Many students have difficulties in traditional geometry courses whenever they are required to construct a proof independently. The requirements of the new course are much more varied: checking a statement for accuracy, confirming correct statements by explanations or proofs, refuting wrong statements by counterexamples, reformulating statements in particular ways, identifying shapes on the basis of givens, and sketching on the basis of givens and known results. This broad variety of question types enables students who have difficulties with proofs to succeed in many exercises and to understand how a mathematical theorem is structured. Their success, in turn, reinforces their motivation to learn geometry.

An evaluation of the new geometry course was carried out in an experimental/control group format. The evaluation included twenty-two experimental classes from fifteen different schools and ten control classes from other schools. Both the experimental and the control classes received two hours of geometry instruction each week for the entire school year and followed the same curriculum topics. The difference was thus solely in the text materials and the mode of instruction.

The evaluation focused on the students' understanding of the deductive reasoning required in Euclidean geometry, on their ability to reason logically, and on their knowledge of geometrical facts. For this purpose three geometry tests and two logic tests were given during the school year. The logic tests contained questions of the type used by Eisenberg and McGinty (1974) and others in studies analyzing students' difficulties with syllogisms. (Sample question: It is known that all basketball players are tall. Dan is not a basketball player. Can you conclude that Dan is not tall?) The geometry tests were carefully designed to test for both factual knowledge of geometry and the ability to reason logically within a geometric context. They were formulated so as not to favor either the experimental or the control group. Sample question:

Given: PAUL is a parallelogram.
         Conclusion: $\angle PLA = \angle UAL$.
         Determine the appropriate explanation:

  a) $\angle PLA$ and $\angle UAL$ are alternate interior
       angles between PA ∥ UL.

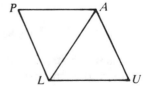

b) Any pair of alternate interior angles are equal.
c) ∠ *PLA* and ∠ *UAL* are alternate interior angles between *PL* ‖ *AU*.
d) △ *UAL* and △ *PLA* are isosceles triangles.

The overall test results are summarized in table 5.1. The statistical analysis showed that the experiment had no effect on the general logical reasoning ability of the students outside a geometric context: there was no difference between the experimental and the control group on the final logic test.

TABLE 5.1

Test Results for Experimental and Control Group

Group	Pretests		Midyear test	Final tests	
	Logic %	Geometry %	Geometry %	Logic %	Geometry %
Experimental	64	60	64	67	60
Control	64	66	53	67	50

The experiment did have, however, a pronounced effect on geometric ability. There were statistically significant differences on both the midyear and the final geometry tests. It was of interest to examine in further detail the source of these differences and in particular their relation to logical reasoning within a geometric context. To get a clearer picture of the cognitive differences between the two groups, the geometry tests were split into two subtests: one containing questions that focused on geometrical knowledge (e.g., computation of angles) and the other containing questions that emphasized the logical aspect (e.g., questions requiring a counterexample). A comparison of student achievement on the two subtests indicates that at midyear the experimental group outperformed the control group by about 12 percent on both subtests, whereas in the final test the advantage of the experimental over the control group was higher on the subtest stressing the logical aspect (18 percent) than on the knowledge subtest (9 percent). It thus appears that the approach taken in the experimental course results in a clear advantage in student ability to reason logically within a geometric context while also improving the geometric knowledge somewhat more than a traditional approach.

Conclusion

It should be apparent that the six logical principles on which this course is based are applicable to any course on Euclidean geometry, independent of its exact content and the mathematical sophistication of the students. Moreover, these principles concern mathematical reasoning in general and may thus be applied to any course or topic with a strong logical component.

One example was given by Halmos (1956), in his book on finite-dimensional vector spaces. He wrote in its preface that if

> an exercise is neither imperative ("prove that . . .") nor interrogative ("is it true that . . .?") but merely declarative, then it is intended as a challenge. For such exercises the reader is asked to discover if the assertion is true or false, prove it if true and construct a counterexample if false. (p. vi)

A more systematic and more widespread acceptance of the spirit behind Halmos's idea would surely lead to a better understanding of the nature of mathematics among the general student population.

REFERENCES

Atiyah, Michael F. "Trends in Pure Mathematics." In *Proceedings of the Third International Congress on Mathematical Education,* edited by Hermann Athen and Heinz Kunle. Karlsruhe, Federal Republic of Germany: The Conference, 1977.

Balacheff, Nicholas. "Experimental Study of Pupils' Treatment of Refutations in a Geometrical Context." In *Proceedings of the Ninth International Conference for the Psychology of Mathematics Education,* edited by Leen Streefland. Utrecht, The Netherlands: State University of Utrecht, 1985.

Deer, George Wendell. "The Effects of Teaching an Explicit Unit on Logic on Students' Ability to Prove Theorems in Geometry." (Doctoral dissertation, Florida State University, 1969.) *Dissertation Abstracts* 303 (1969): 2284 (University Microfilms 69-17,669).

Eisenberg, Theodore A., and Robert M. McGinty. "On Comparing Error Patterns and the Effect of Maturation in a Unit on Sentential Logic." *Journal for Research in Mathematics Education* 5 (November 1974): 225–37.

Emerson, W. *The Elements of Geometry.* In which, The principal Propositions of Euclid, Archimedes, and others, are demonstrated after the most easy manner. London: J. Nourse, 1763.

Fischbein, Efraim, and Irith Kedem. "Proof and Certitude in the Development of Mathematical Thinking." In *Proceedings of the Sixth International Conference for the Psychology of Mathematics Education,* edited by Alfred Vermandel. Antwerp: The Conference, 1982.

Fletcher, T. J. "The Teaching of Geometry: Present Problems and Future Aims." *Educational Studies in Mathematics* 3 (1970–71): 395–412.

Freudenthal, Hans. *Mathematics as an Educational Task.* Dordrecht, The Netherlands: Reidel, 1973.

Friedlander, Alex, and Nurit Hadas. *Chapters in Plane Geometry* (in Hebrew). Rehovot, Israel: Weizmann Institute of Science, 1981.

Halmos, Paul R. *Finite-dimensional Vector Spaces.* 2d ed. Princeton, N.J.: D. Van Nostrand, 1956.

Senk, Sharon L. "How Well Do Students Write Geometry Proofs?" *Mathematics Teacher* 78 (September 1985): 448–56.

Thom, Rene. "Modern Math—Does It Exist?" In *Proceedings of the Second International Congress on Mathematical Education,* edited by Albert Geoffrey Howson. London: Cambridge University Press, 1973.

Vinner, Shlomo, and Rina Hershkowitz. "On Concept Formation in Geometry." *Zentralblatt fuer Didaktik der Mathematik* 15 (February 1983): 20–25.

6

Geometry: An Avenue for Teaching Problem Solving in Grades K–9

Linda J. DeGuire

MANY reasons can be given for studying geometry in the elementary and middle grades. One of these is the opportunity geometry affords both to teach problem solving and to teach for problem solving. Let me clarify these last two phrases: "Teaching problem solving" goes beyond merely solving problems to include reflecting on the solution processes in order to glean problem-solving strategies that might be useful in later problem-solving episodes. "Teaching for problem solving" involves teaching content in a meaningful way so that it can be used in further problems and learning. At least one way to teach for problem solving is to have the content develop from problem-solving episodes. This article will give examples of using selected geometric activities to teach problem solving and to teach for problem solving on three levels—early childhood, intermediate, and middle grades. Though the examples are categorized by level, they can be simplified or extended for use at the other levels.

Patterns—a Basic Tool in Problem Solving
(Early Childhood Level, K–3)

Some mathematicians would define all mathematics as seeing patterns. Whether you agree with that philosophy or not, looking for patterns is an extremely useful and widespread problem-solving strategy. Although some numerical patterns are difficult for young children, visual patterns are well within their reach. Initially, strings of variously colored, interlocking cubes can be used to make simple linear patterns. Later, pattern blocks can be used to make more complex linear patterns and two-dimensional patterns.

Children can begin by copying patterns. Have several patterns ready for them, each shown in a cube string. For example, have the children copy the cube strings in figures 6.1–6.4. At this point, the child is only imitating the arrangement of cubes and may not necessarily discern the underlying pattern.

Many children will be aware of the patterns in the cubes before they can formally describe exactly what is repeated to make the pattern. Activities

Fig. 6.1. Red-white-red-white-red-white

Fig. 6.2. Red-white-white-red-white-white-red-white-white

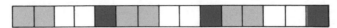

Fig. 6.3. Pink-pink-white-white-red-pink-pink-white-white-red-pink-pink-white-white-red

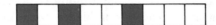

Fig. 6.4. Red-white-red-white-white-red-white-white-white

to help them express or develop this awareness include modifying the pattern and extending the pattern. A child can modify the pattern in figure 6.2 by changing the colors to white-blue-blue-white-blue-blue. . . . Or the same pattern can be extended by adding a cube of a third color, followed by two more white cubes, on the cube string. With either activity, we can clearly see that the child discerns the pattern even when unable to verbalize it.

At some point, we want the children to be able to discriminate between a cube string that presents a pattern and one that does not. To be a pattern, there must be some form of repetition, some way to say what comes next on the basis of what has preceded. In simple patterns like those shown in figures 6.1–6.3 the cube strings can be physically broken into substrings that are identical. Young children can call the substrings "small strings" or "small towers." For example, red-white-white-red-white-white . . . in figure 6.2 can be broken into three identical small strings of red-white-white. In a pattern such as that in figure 6.4, the cube string cannot be physically broken into substrings that are identical, but it can be broken into substrings that are related in such a way that we know what the next substring should be to continue the pattern. That is, it can be broken into three substrings—R-W, R-W-W, and R-W-W-W—that suggest that the next substring should be R-W-W-W-W. However, a string such as R-W-B-G-R-W-W-B does not show a pattern, because it cannot be broken into identical or related substrings. (Of course, it could be the first statement of a pattern that repeats the entire sequence R-W-B-G-R-W-W-B. However, the usual convention for showing patterns in mathematics is to repeat the pattern at least three times.) Such a concrete way to discriminate examples and nonexamples of patterns makes pattern work with colored cubes an ideal activity for young children. No words or numbers are needed.

Once children can discriminate examples and nonexamples of patterns among cube strings, they are ready for the real fun—creating their own patterns. At first, they may be reluctant to move beyond the examples used by the teacher. However, leading questions can help them broaden their search. For example:

- Take some red cubes and some blue cubes. *[Later, add another color.]* Make a pattern with them. Can you make a different pattern with them? Remember to include three or more "small strings" in your "big string."

- Look at this pattern. [Point to one displayed in the classroom somewhere.] Can you make a new pattern by changing the colors?

- Take twelve cubes, six each of two different colors, and make a pattern with them. Can you make another pattern with them?

Toward the end of the early childhood level, children can begin working with more complex or two-dimensional patterns. Pattern blocks (collections of colored plastic or wooden shapes that are related to each other by size) offer a rich variety of two-dimensional patterns. Again, children can copy the patterns, extend them, and create their own. Many such patterns use point and line symmetry (see fig. 6.5). Symmetry in concrete situations can lead to a useful and powerful problem-solving strategy in later mathematics.

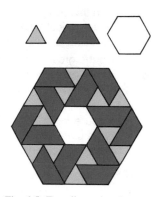

Fig. 6.5. Two-dimensional pattern

After the children can discriminate patterns and nonpatterns, they can be given certain conditions and challenged to fulfill them in as many ways as possible. The tasks in figures 6.6 and 6.7 suggest some pattern block activities appropriate for elementary children. Such activities are excellent beginnings for important problem-solving strategies such as "make an organized list," "list all possibilities," "vary the conditions of the problem," and "look for patterns."

Activities on the Geoboard
Intermediate Level, 4–6

Children on the intermediate level have usually reached a level of cognitive development that enables them to reason deductively with objects that

Tasks:

a) Fill in this design—
- with just two blocks;
- with exactly three blocks;
- with exactly four blocks.

b) What is the maximum number of blocks you can use to fill in the design?

c) Can you fill the design with five blocks, six blocks, and so forth [up to the maximum number]?

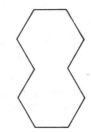

Fig. 6.6. Problem-solving tasks
with pattern blocks

Tasks:

a) Fill in this design—
- using just 2 colors of blocks
- using just 3 colors.

b) Use the green triangle as the unit of measure. What is the area of this figure? Make a different figure with this same area.

c) Now let the blue rhombus be the unit of measure for the area. Try to find the area in blue units *without* laying blue units on the figure and counting.

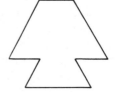

Fig. 6.7. Problem-solving tasks
with pattern blocks

are actually present or with which they have had recent experience. Many geometric manipulative aids provide opportunities for reasoning with objects and thus for teaching problem solving and teaching for problem solving. One such aid is the geoboard. Results from geoboard activities can be recorded on dot paper, preferably with the dots arranged to correspond to the actual geoboards being used. There are many activities on the geoboard from which children can derive enjoyment and benefit. For example, they can make figures on the boards, name figures already made, count to find areas and perimeters of figures, and use the nails as a coordinate or grid system. Each of these can be an excellent learning activity and can also be extended to provide problem-solving challenges.

Extensions of Making Figures

Consider the task of making given figures on the geoboard. The teacher might use such a task to reinforce the names and characteristics of figures such as triangles or parallelograms. Students could also be asked to make less traditional figures, such as a figure with exactly two right angles, or a figure that leaves exactly five nails outside of it, or a figure with an area of six square units. More than one condition can be specified so that the task is challenging even to the brightest students. To complete the task, children can be encouraged to call on a variety of problem-solving strategies, such as trial and error, attention to one condition at a time, the coordination of several conditions at a time, and looking back to insure that all conditions have been met.

The task of making given figures on the geoboard can also be extended to the task of finding all possible figures that meet certain conditions. For example, the task of making a figure with a perimeter of ten units can be extended to finding all possible figures with a perimeter of ten units. Such an extension calls into play other important problem-solving strategies, such as systematic trial and error, patterns in ways to search for and record possibilities, and methods to insure that all possibilities have been considered and found.

Tasks of making figures to meet certain conditions can also be used to teach for problem solving. For example, certain number attributes (such as even and odd, prime and composite, square and nonsquare) and number relationships (such as factor, divisor, and multiple) can be drawn from, and exemplified with, tasks on the geoboard. Children can be given the task of making rectangles so that one side is length 2 and then finding the area of each of their rectangles by counting the squares inside it. The task can be extended to the following problem:

> Given a whole number, can you always make a rectangle on the geoboard with one side of length 2 that has the given number as its area? If so, why? If not, why not? With what whole numbers can you do so? How can you tell ahead of time whether or not you will be able to make such a rectangle with its area a given whole number?

A child working with such a problem task might initially call on the problem-solving strategy of "guess and test." The testing may involve an attempt to "consider all possibilities" or "make an organized list." Attempts to express the hypothesis may lead the child to "make a chart or table" and "search for a pattern" (i.e., every other number). The child may feel confident of his or her hypothesis on the basis of examples. Then the teacher can direct the child to use deductive reasoning to relate the result to other known facts, perhaps multiplication facts or divisors of the numbers. Of course, experience with such a problem task can serve as a basis for the concepts of even and odd numbers. Such a visual basis will give the child richer concepts than just the numerical point of view (i.e., divisible or not by 2). Thus, the problem above is an example of a task through which we can teach problem solving and also teach for problem solving.

Another example of using geoboard figure activities both to teach problem solving and to teach for problem solving can be seen in extensions of the task "Make a rectangle that has an area of 8." An initial extension of the task is to ask, "In how many different ways can you make a rectangle with an area of 8 square units?" A child working on such a problem may decide to make a chart by recording the results in a 2-by-4 form, with the two numbers indicating the length and width of the rectangle. A child can also decide to explore all possibilities, preferably in a systematic method, by first considering all 1-by-? rectangles, then all 2-by-? rectangles, then all 3-by-? rectangles, and so on, until all 8-by-? rectangles have been considered. This initial extension of the original task provides a basis for the following problems, which are further extensions of the original task:

- Given any whole number, in how many ways can you make a rectangle with an area equal to the given whole number of square units? Can you answer this question without actually drawing or making rectangles? How?

- What kinds of numbers do you get for the number of ways to make rectangles with a certain area? Does the kind of number you get for a certain area relate to the kind of number given for the area? How?

(If you have not worked with these tasks before, I strongly encourage you to pull out a geoboard or dot paper and actively engage in the tasks before reading further. Otherwise, you'll spoil your own fun.) The following problems call on not only the strategies of making an organized list and recording the results in some meaningful way but also the strategies of looking for a pattern, making a hypothesis and testing it, and using deductive reasoning. Of course, the problem-solving strategies used in the solution attempt can also be used to teach problem solving. Also, the results can be used to give visual representation to the concepts of *prime number* (there is only one way to make a rectangle with the given area), *composite number* (there is more than one way to make a rectangle with the given area), and *square number* (there are an odd number of ways to make a rectangle with that area; a perfect square can be made with the given number as area).

Guess the Figure

Another problem-solving activity involving geometry and the geoboard is "guess the figure." In this activity, a leader (the teacher or a student) makes a figure on the geoboard and does not show it to the class. The class must then try to identify the figure by asking questions that can be answered by yes or no. Various means can be used to make the final guess. One way is to have children make a copy of the figure on their own geoboards and turn them face down. Then the teacher can quietly check the corrections of the guesses without spoiling the fun for students who haven't yet figured it out. Many problem-solving strategies must be called on to guess the figure correctly. These strategies can be explicitly discussed and refined after the game. The following dialogue represents part of a possible game episode in a sixth-grade class.

Leader: OK. I've got a figure ready on my geoboard. Ask me questions to try to figure out what it is.

Ann: Does it have any right angles? (*Leader:* Yes.)

Mike: How many right angles?

Teacher: Oops, Mike. Your question can't be answered by yes or no. So your team loses one point.

Bob: Does it have at least three sides? (*L:* Yes.)

Karen: I'd like to make a guess. Is it a right triangle? (*L:* No.)

Teacher: Sorry, Karen. Your team loses one point.

Mike: Does it have more than one right angle? (*L:* Yes.)

Janet: Does it have four right angles? (*L:* No.)

Bob: Does it have two right angles? (*L:* Yes.)

Ann: Does it have more than three sides? (*L:* Yes.)

Karen: Does it have exactly four sides? (*L:* Yes.)

Mike: Are all pairs of opposite sides parallel? (*L:* No.)

Ann: Is one pair of opposite sides parallel? (*L:* Yes.)

Janet: I'd like to make a guess. Is it a . . .?

(If you're stumped, the answer is in figure 6.10 at the end of this article.)

The activity above can be greatly complicated by imposing a grid system on the geoboard and extending the task to include guessing the exact location and size of the figure. Let the lower left corner of the geoboard be the origin. Thus, the nails or pegs are at the points (1, 1) through (5, 5). For the figure in the game above, answers to the following questions could be used to help determine its location and size.

- Does each side touch at least three nails? (No.)
- Is one corner at position (1, 2)? (Yes.)
- Are all the sides parallel to a row or column of nails? (No.)
- Does it have at least one nail completely inside it? (No.)

The guess-the-figure activity encourages the children to use language carefully (e.g., the difference between "at least three sides" and "three sides") and to reason deductively from the information obtained from each question and answer. Much of the reasoning takes the form of considering all possibilities and eliminating certain ones, very useful problem-solving strategies that should be discussed explicitly when reflecting back on a particular episode.

Can You Solve It Another Way?
(Middle Grades Level, 6–9)

In the looking-back stage of problem solving, the strategy "Can you solve it another way?" is useful in strengthening or weakening one's confidence in the solution to a problem. Elementary school children frequently do not yet have available the cognitive and mathematical tools to use the strategy effectively. However, many students in the middle grades have the cognitive maturity to begin finding or understanding alternative, deductive solutions to problems. The visual aspects of many geometric problems aid their ability to deal with the formal deductive reasoning.

A good example of a geometric problem that can be solved in more than one way by students in the middle grades is to find the sum of the interior angles of a convex polygon. Initially, students might measure and add the measures of the interior angles, or they might decompose the figure into triangles and count the triangles. At this point, students would be encouraged to "do a simpler problem" (i.e., find the sum for a three-sided polygon, a four-sided polygon, a five-sided polygon, etc.), "make a chart," and "look for a pattern." After the pattern is found, students could be led to prove the result informally but deductively.

Although the problem above is contained in many middle grades mathematics textbooks, a modification of that problem affords a surprising result and an excellent opportunity for an alternative, deductive solution. The modified problem is to find the sum of the exterior angles of a convex polygon. (Once again, if the reader has not previously worked with this problem, I strongly encourage you to attempt it before reading further.) If the students have previously solved the sum-of-the-interior-angles problem, I suggest that you merely present the sum-of-the-exterior-angles problem to them without many hints or directions. You may need to review what an exterior angle is, and you may want to clarify that since both exterior angles at a vertex have the same measure, only one will be included from each vertex. Many students will begin by drawing one or more convex polygons and measuring their exterior angles. Even if most middle grades students were adept at using protractors, a certain amount of error is inherent in the measurement process. Results from adding the measures of the exterior angles of the polygons that were drawn will be tentative at best. Some students may begin with, or progress to, "doing a simpler problem," "making a chart," and "looking for a pattern." This approach may increase the students' confidence that the result is always 360 degrees. But, once again, the error inherent in the measurement process and in the students' use of the protractor will probably leave some doubt about the conclusion.

The formal, deductive proof of the result that the sum of the exterior angles of a convex polygon is always 360 degrees depends on mathematical induction and other mathematical results not yet accessible to the student in the middle grades. However, the visual nature of the problem allows most students at this level to understand an informal presentation of the deductive proof. In fact, I have even had bright eighth-grade students arrive at such a solution on their own. The following is such an informal, deductive explanation. It uses

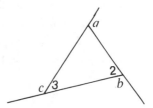

Fig. 6.8. Triangle with exterior angles marked

the problem-solving strategies of "do a simpler problem," "make a chart," "draw a diagram," and "look for a pattern." See figure 6.8.

Begin with a triangle. Label its exterior angles with lowercase letters and its interior angles with numbers. Note that each exterior angle is supplementary to the adjacent interior angle. Thus we have three equations:

$$m\angle a + m\angle 1 = 180$$
$$m\angle b + m\angle 2 = 180$$
$$m\angle c + m\angle 3 = 180$$

Add the three equations together and we have

$$m\angle a + m\angle b + m\angle c + m\angle 1 + m\angle 2 + m\angle 3 = 3(180).$$

But $m\angle 1 + m\angle 2 + m\angle 3 = 180$. After substituting this relationship into the equation above, we have

$$m\angle a + m\angle b + m\angle c + 1(180) = 3(180).$$

Finally, subtract 1(180) from both sides of the equation and—voilà!—we have the sum of the exterior angles:

$$m\angle a + m\angle b + m\angle c = 2(180), \text{ or } 360.$$

It would be wise to repeat this explanation with at least one more triangle that is a different shape from the original. It is important for the students to realize that the equations above depend only on the fact that the figure is a triangle, not on any characteristics of a particular triangle or kind of triangle.

The procedure can then be repeated for a quadrilateral, pentagon, and perhaps a hexagon. It is repeated here for the pentagon only in order to illustrate the generality of the procedure. Draw any pentagon; avoid drawing a regular pentagon. Label its exterior angles with lowercase letters and its interior angles with numbers (see fig. 6.9). Note that each exterior angle

is supplementary to the adjacent interior angle. So, we have the following
five equations:

$$m\angle a + m\angle 1 = 180$$
$$m\angle b + m\angle 2 = 180$$
$$m\angle c + m\angle 3 = 180$$
$$m\angle d + m\angle 4 = 180$$
$$m\angle e + m\angle 5 = 180$$

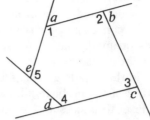

Fig. 6.9. Pentagon with exterior angles marked

Now add the five equations together, and we have

$$m\angle a + m\angle b + m\angle c + m\angle d + m\angle e + m\angle 1 + m\angle 2 + m\angle 3 + m\angle 4 + m\angle 5 = 5(180).$$

But the sum of the measures of the interior angles ($m\angle 1 + m\angle 2 + m\angle 3 + m\angle 4 + m\angle 5$) is 3(180). After substituting this relationship into the
equation above, we have

$$m\angle a + m\angle b + m\angle c + m\angle d + m\angle e + 3(180) = 5(180).$$

(Notice that no matter how many sides the polygon has, the sum of the
interior angles is always $(n - 2)(180)$ and the other side of the equation is
always $n(180)$.) Once again, subtract 3(180) from both sides of the equation
to have the sum of the exterior angles:

$$m\angle a + m\angle b + m\angle c + m\angle d + m\angle e = 2(180), \text{ or } 360$$

Problems such as the one just discussed give students in the middle grades
a chance to use their developing deductive reasoning powers and problem-
solving strategies with visual objects. The union of the two abilities, whether
working with the actual objects or drawings of the objects, will strengthen
the development of each.

Conclusion

Geometry is an excellent source of problem-
solving episodes, and I have selected only a few.
I hope readers will be convinced of the usefulness of geometry for problem solving and will
be motivated to explore the area further in order to build up their own repertoire of such
episodes.

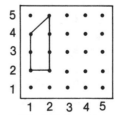

Fig. 6.10. Solution to
Guess-the-Figure game

7

Creative Geometry Problems Can Lead to Creative Problem Solvers

George A. Milauskas

IT IS my contention that students learn problem solving by solving quality problems. The blend of drill and exposure to nonstandard problems is what stimulates students to exercise their faculties for problem solving. An emphasis on problem solving does not consist in merely inserting a few "specialty problems" here and there in the classwork. Rather, problem solving should be the underlying theme in the mathematics classroom. Each assignment should incorporate problems designed to encourage flexibility and thought. Problems should use skills from other disciplines as well as encourage the learning of sound problem-solving techniques and the use of higher-order thinking skills. Mathematics becomes more meaningful to the student who is routinely exposed to variety in problem sets. Such a student will be better able to adapt to new situations and to approach new problems with confidence.

My goal here is to inspire you, the teacher, to go out of your way to stimulate your geometry class with problems that take the students beyond rote drill. Some creative problems are included for your use in the classroom as well as a few that may entertain and challenge you too.

Types of Problems

Let's take a look at a variety of problems that vary in their level of sophistication and that exercise different problem-solving techniques.

Recognition

Find and count all the squares in each figure.

Find and count all the rectangles in each figure.

a)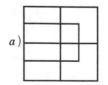

b)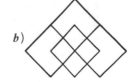

Grateful acknowledgment is given to the University of Wisconsin Mathematical, Engineering, and Science Talent Search and to the Illinois Council of Teachers of Mathematics State Contest Committee for providing some of the problems used in this article. Additional sources of problems are listed in the Bibliography.

Basic Drill and Algorithm Practice

a) Find the hypotenuse of a right triangle whose legs are 6 and 8.

b) Find the height of an isosceles triangle whose legs are each 17 and whose base is 16.

c) Find all right triangles whose legs are chosen from the numbers 3, 4, 6, and 8. (Consider all combinations including repeats, such as 3 and 3.)

d) Find all "triples" (combinations of three whole numbers) that can be the lengths of the sides of right triangles with a hypotenuse less than 30.

Notice the difference between drill problems (*a*) and (*b*) and drill problems (*c*) and (*d*); the latter are somewhat "open."

Applications

For students at higher learning levels these application problems, admittedly contrived, may not be much more than basic drill. For others, they can be the next important step toward developing problem-solving skills.

a) A 17-foot ladder is leaning against a house. The base of the ladder is 9 feet from the house. How far up the wall does the ladder reach?

b) A kite is flying at the end of 100 feet of string that extends at an angle of 60 degrees to the ground. How far above the ground is the kite?

c) Find all right triangles whose sides are consecutive integers. (Let the sides be "x," "$x + 1$," and "$x + 2$.")

Open Applications

True problem solving, in the eyes of some educators, does not take place until we get to the stage where a strategy is not clearly evident in the wording of the problem.

a) Some students found a pirate treasure map with the following clues: "Take ye 20 paces East of the old oak tree; then 15 North and 18 West. Walk 9 paces North and another 5 East and here ye find me treasure." How many paces, in a straight line, was the treasure from the oak tree? [Answer: 25] (*Hint:* Form a single right triangle by combining all east-west motions and all north-south motions into single values.)

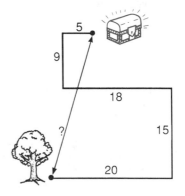

b) A boat is tied by a 20-foot rope to a point on a pier, 12 feet above water level. If 5 feet of rope are pulled in, how far forward does the boat move? [Answer: 7 feet]

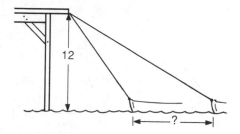

True Applications

True application problems can be obtained through interdepartmental work with science, technology, and other areas or through personal experience. The first example, the hydraulic piston problem, is from technology. The second is a problem I had to solve to set up my son's swimming pool.

a) See figure 7.1.

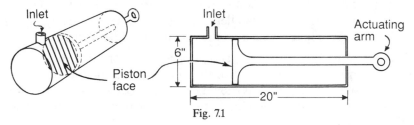

Fig. 7.1

1. Find the area of the piston face.
2. Find the volume of fluid needed to push the piston 18 inches.
3. If fluid enters at 2500 psi (pounds per square inch), what is the force on the actuating rod?

b) Andy's round swimming pool is 6 feet in diameter. Chemical powder must be mixed with the water at a rate of one packet per 200 gallons of water. If the water is 2 feet deep, how many packets are needed?

Algebra

In addition to enhancing problem-solving skills, some geometry problems ought to serve as a vehicle for reviewing other skills, like algebra. Algebra and geometry are frequently thought of as disjoint subjects, but creative problems can provide a link between these seemingly disparate worlds.

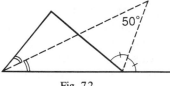

Fig. 7.2

An interior and exterior angle of two distinct angles of a triangle are each bisected, as shown in figure 7.2. If the angle formed by the bisectors is 50°, find the degree measure of the third angle of the triangle. The solution can be found by using an applica-

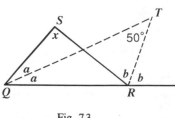

Fig. 7.3

tion of algebra (see fig. 7.3): Since $2b = 2a + x$ (exterior angle of $\triangle QRS$), $x = 2b - 2a = 2(b - a)$. But in $\triangle QRT$, $a + 50 = b$; so $b - a = 50$. Thus $x = 2(b - a) = 2(50) = 100°$. Notice not only the application of algebra but also that neither a nor b can be determined. The "dual substitution" of 50 for $b - a$ is an enhancement of skills developed in algebra.

Extensions

a) What if we could discover an exterior angle theorem for quadrilaterals? What relationship would we find?

b) A trapezoid may be defined to be "a quadrilateral having *exactly* one pair of opposite sides parallel" or "a qualrilateral having *at least* one pair of opposite sides parallel." Compare these two definitions and discuss the impact of each on the quadrilateral family tree.

The question "What if . . ." is an important one for the geometry classroom. Extensions allow for creative thought approaching the analysis and synthesis levels. Some problems should encourage conjecture and guessing. Students must be exposed gradually to problems that incorporate aspects of higher-order skills.

Open Searches

a) I have two sticks. One is 5 cm and the other 12 cm long. What are all the possible lengths (in whole centimeters) of a third stick I could use with the others to construct a triangle?

b) Find all triangles with integer sides that have a longest side 10.

c) Consider all rectangles with integer sides and perimeter 12; which has the greatest area?

Hints on Creating and Using Quality Problems

As you set out to create good problems for your geometry students, there are certain things to keep in mind. Try to find problems that are simple to state but that have a twist or a novel solution. Realistic problems can be motivating, but some that are totally unreal, unexpected, or unusual can also prove motivating. Such problems pique curiosity and invite solution. Good problems sometimes contain insufficient or extraneous information, so that the student has to think in terms of necessary and sufficient conditions. Sometimes it is not the problem itself that is important but rather the thought, analysis, and techniques necessary for its solution.

One should also consider the way a problem is posed. The way a question is asked may suggest or limit the strategy used to solve the problem. It may have an effect on the motivation to spend time on the problem and may have a bearing on the student's ability to solve the problem at all.

The teacher needs to exercise control over how and when a problem is used. Hints or lead-in activities may be needed. Perhaps students should be allowed to work in groups. A problem may be more suitable for classroom discussion than for use in a homework set.

Encourage alternative solutions and other extensions offered by students. The best solutions are those that are general, so that they can be applied to future problems. What I find most exciting about teaching is the enthusiasm and pride that students exhibit when they are inspired to use a technique that they have abstracted from earlier work. Such creative insights are not solely the domain of the better student. Many "average" students will come up with unusual ideas worthy of further discussion.

It is certainly advisable for you to work the problems in this article before using them in the classroom. This will enable you to explore some blind alleys, as a student might, as well as gauge the problem's level of difficulty. Such a preview may provide a fuller understanding of the problem and allow for a solution that is different or even more elegant than the one given.

Choose problems appropriate to the topic and to the ability of the student. Some problems included here may be familiar to you, but I hope many will not. Often the beauty of the problem lies in a novel or alternative solution rather than in the problem itself. Hints and answers follow the problems. I hope you find the problems entertaining, educational, thought-provoking, and most of all, stimulating to your own creative ability to pose problems for your students. I further hope that you feel so strongly about the emphasis on problem solving that you will work to become an active problem solver as well as an active teacher of problem solving.

Problem Set

1. Two congruent 6 cm × 6 cm squares overlap as shown. A vertex of one square is at the center of the other square. What is the largest possible value for the colored area? (The top square is movable, as long as the vertex remains in the center, as shown.)

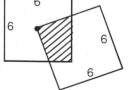

2. *a*) A 6-by-8-inch paper rectangle is folded so that opposite vertices touch. Find the length of the fold.

b) An 8-by-12-inch rectangle is folded so that a vertex touches the midpoint of the longer nonadjacent side. Find the length of the fold.

3. Given a unit square, find the area of interior square *S*, which is determined by joining each vertex of the unit square to the midpoint of its clockwise nonadjacent side.

4. A touch of calculus?

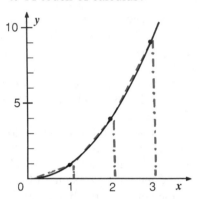

a) Estimate the area under the curve $y = x^2$, $0 \le x \le 3$, by adding the areas of the two trapezoids and the triangle shown.

b) Estimate the area under the curve using 1/2 unit *x*-intervals and adding areas of five trapezoids and a triangle.

c) How could we get a better estimate?

5. Each edge of the cube shown is 6 cm. Find the distance from the midpoint of one of the face diagonals to the furthest vertex of the cube.

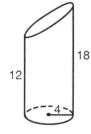

6. A cylinder is cut on a slant as shown. The height goes from 12 cm to 18 cm, and the radius is 4 cm.

a) Find the volume.

b) Find the lateral area.

c) Find the total area. (*Note:* The area of an ellipse is $\pi R r$.)

7. Given isosceles triangle *AEF* (*AF = AE*) with a path of five congruent segments *A-B-C-D-E-F*. Find the degree measure of angle *A*.

Extension: Analyze the problem for a path of 3 congruent seg-

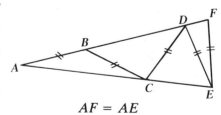

$AF = AE$

ments; 4 congruent segments; 6 congruent segments; n congruent segments.

8. *a*) Which convex polygons, from triangle through duodecagon, can be constructed from squares and equilateral triangles? (Demonstrate.)

   b) Prove that a convex plane figure that can be divided into squares and equilateral triangles cannot have more than 12 sides.

9. In a certain parallelogram $ABCD$, the bisectors of two consecutive angles (A and D) meet at a point (P) on a nonadjacent side. Tell all you can about the sides, angles, and triangles.

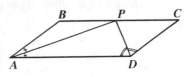

10. Given a quadrilateral that has one pair of opposite sides congruent, the other pair not congruent, and a pair of opposite angles that are supplementary. Prove that this quadrilateral is an isosceles trapezoid. (Challenging)

11. If a point is selected at random in the interior of a circle, find the probability that the point is closer to the center than to the circle.

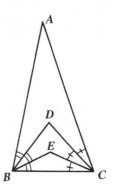

12. The corresponding trisectors of two angles (B and C) of a scalene triangle meet at points D and E. The third angle of the triangle (angle A) is 30 degrees. Find the measures of angles D and E.

13. The bisectors of the upper base angles (R and A) of a trapezoid meet at point B. Find the relationship between the measures of the lower base angles (T and P) and angle B.

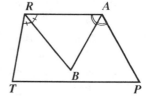

14. An equilateral triangle is inscribed in a circle and another equilateral triangle is circumscribed about the circle. Find the ratio of the areas of the triangles.

15. Given the following patterns formed by joining (*) four simple patterns, A, B, C, and D:

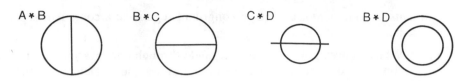

A * B   B * C   C * D   B * D

Which of the following is $A*D$? $A*C$? (Justify your answer.)

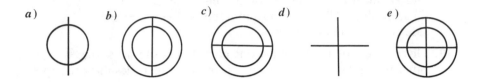

a)   b)   c)   d)   e)

16. A rectangle is divided into four rectangles with areas 45, 25, 15, and x. Find x.

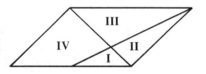

17. The area of a parallelogram is 60. A segment is drawn from one vertex to the midpoint of an opposite side. The diagonal is drawn between two other vertices as shown. Find the areas of the four regions formed.

18. a) Find the area of a triangle with vertices $(-5,7)$, $(-2,3)$, and $(6,-1)$.

b) Find the area of a quadrilateral with vertices $(-5,7)$, $(-2,3)$, $(6,-1)$, and $(11,-1)$.

c) What is the most descriptive name for the figure described in part (b)?

d) *Prove:* If the coordinates of the vertices of a polygon are rational, then the area is rational.

19. Given triangle *ABC:* $A(-3,-10)$, $B(-12,2)$, $C(12,10)$. Find the point where the bisector of angle A intersects side BC.

20. Find the perimeter of a rectangle whose area is 22 square meters and whose diagonal is 10 meters.

21. Given a square divided into two smaller rectangles (shaded) and two squares, $TR = 7$ and $UE = 20$.

    a) Find PQ.

    b) Find the total shaded area.

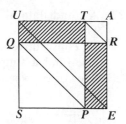

22. What figure is formed when the consecutive midpoints of the sides of a quadrilateral are joined? (*Note:* The result is the same whether the quadrilateral is convex, concave, simple or not simple, or even noncoplanar.) What if the original quadrilateral were a rectangle? A kite? An isosceles trapezoid? A square? A rhombus?

23. The perimeter of a rectangle is 48. Create a table to display the possible integer lengths and widths of such a rectangle and the area of each.

    a) What are the possible values of the width? The length?

    b) What is the range of areas you found? The maximum area?

    c) Analyze such a table for a rectangle where the sum of *three* sides is 48, and determine the range of areas. Find the maximum area.

24. The area of a rectangle is 120. Find the perimeter if the width is 1, 2, 3, 4, ... (Make a table showing width, length, and perimeter for each.) What do you notice from the data? What extensions are possible?

25. A circular lighthouse floor has a circular rug in the center. The lighthouse keeper observes that if he places a 10-foot pole on the floor so that each end touches the wall, then it is tangent to the rug. Find the area of the floor that is *not* covered by the rug.

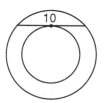

26. a) The sides of a triangle are 6, 8, and 10. Find the radius of an inscribed circle.

    b) The centers of three mutually tangent circles are the vertices of a triangle with sides 6, 8, and 10. Find the radius of the smallest circle.

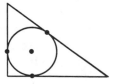
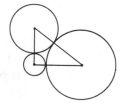

27. A quadrilateral is circumscribed about a circle. Three of the sides of the quadrilateral are 9, 17, and 12, consecutively. Find the length of the fourth side (x). Find a simple relationship between the four sides of every such circumscribed quadrilateral.

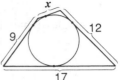

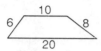

28. Find the area of a trapezoid whose bases are 10 and 20 and whose legs are 6 and 8.

29. Given three points, *A, B,* and *C,* on an ellipse with major axis 10 and minor axis 8. The foci are *F* and *G.* The sum of the distances from *F* to *A, B,* and *C* is 12 (that is, *FA* + *FB* + *FC* = 12). Find the sum of the distances from *G* to *A, B,* and *C.*

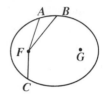

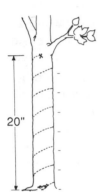

30. A nuthatch pecks its way around a tree trunk starting from a point 20 feet overhead and working down to the ground. The bird follows a spiral path (helix) that takes it seven times around the 3-foot circumference of the tree. Find the total distance traveled by the nuthatch.

31. Find the set of convex polygons for which the number of diagonals is greater than the sum of the measures of the interior angles.

32. A path travels from point *P* parallel to consecutive sides of a scalene triangle as shown. What will happen if the path continues in this fashion? Prove your conjecture.

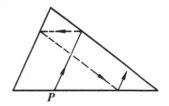

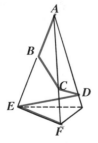

33. A bug crawls from the apex of a regular pyramid around the figure along a path of equal segments, *AB* = *BC* = *CD* = *DE* = *EF.* Find the measure of the vertex angle of each of the three identical isosceles trianglar faces. (If you need a hint, go back to problem 7.)

34. Triangle *A* is cut from cardboard that is white on one side and black on the other. Place triangle *A* in the position shown with the white side up. Show how to cut the triangle into pieces that can be reassembled into the reflected congruent triangle *B* so that the top is still white.

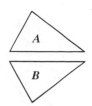

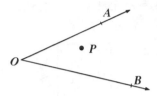

35. Given an angle *AOB* and *any* point *P* in its interior. Use straightedge and compass to construct a segment $\overline{XY}$, with *X* on $\overline{AO}$ and *Y* on $\overline{OB}$ so that *P* is the midpoint of $\overline{XY}$.

36. Prove that in a triangle with sides 4 cm, 5 cm, and 6 cm one of the angles is exactly twice as large as another.

37. We know the median of a trapezoid is the average of the bases. Find the lengths of the segments that join the trisection points of the legs of a trapezoid and are parallel to the bases.

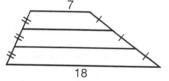

38. Draw two lines from the same vertex of a parallelogram that will divide the figure into three regions of equal area.

Hints and Solutions

1. Move one square relative to the other and observe: $S = 1/4$ square, so the shaded area is always 9.

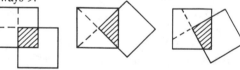

2. Note that the fold is the perpendicular bisector of the line joining the vertices. Try coordinate geometry for a change, or use similar triangles.

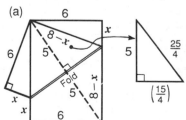

(a)

$$x^2 + 6^2 = (8-x)^2$$
So $x = \frac{7}{4}$
$$8 - x = \frac{25}{4}$$

(b)

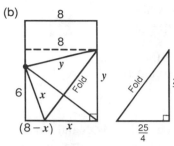

$$6^2 + (8-x)^2 = x^2 \qquad x = \frac{25}{4}$$

$$(y-6)^2 + 8^2 = y^2 \qquad y = \frac{25}{3}$$

Answers: a) 15/2 b) 125/12

3. Observe that $A + B = S$, so the square $= 5S$, or $S = 1/5$

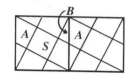

4. *a*) 9 1/2 *b*) 9 1/8 *c*) With smaller intervals, more trapezoids. (*Note:* the actual area, by integral calculus, is 9.)

5. Draw two more face diagonals to form an equilateral triangle with side $6\sqrt{2}$. So the required segment is $3\sqrt{6}$.

6. Here are some of the many solutions that students have suggested:

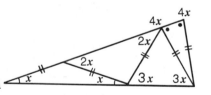

Answer: $V = 240\pi$ $LA = 120\pi$ $TA = 156\pi$

7. Repeating applications of the exterior angle theorem and solving

$$4x + 4x + x = 180,$$

we find that the angle at A is 20 degrees. *Extension:* For N segments, $m\angle A$ is $180/(2N - 1)$.

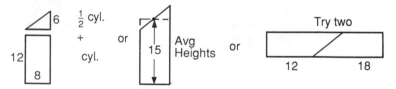

8. *a*) Some of the harder figures:

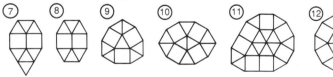

b) A convex plane figure that can be so constructed would have interior angles of 60°, 90°, 120°, or 150°. Thus, each exterior angle must be at least 30°. There can be, at most, $360/30 = 12$ sides.

9. Triangles ABP and CDP are isosceles; angle APD is a right angle; $BC = 2\,AB$; and P is a midpoint of $\overline{BC}$. Also, triangles ABP and CDP have areas each equal to a fourth of the parallelogram.

10. Don't feel bad if you didn't get this one! Any quadrilateral with opposite angles supplementary can be inscribed in a circle. *Opposite sides equal* yields a pair of equal arcs, each intercepted by equal inscribed angles. One pair of opposite sides are parallel. The second pair of opposite sides are congruent.

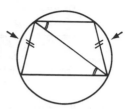

11. Probability $= \dfrac{\text{(area of } r/2 \text{ circle)}}{\text{(area whole)}} = 1/4$

12. Let $3X$ and $3Y$ be the trisected angles. $3X + 3Y + 30 = 180$; thus, $X + Y = 50$ and $D = 180 - (2X + 2Y) = 80$ and $E = 180 - 50 = 130$.

13. A similar argument to problem 12 yields: Angle B is the average of the lower base angles of the trapezoid.

14. "Cut out" along the circle and rotate the interior region $180°$ to create an equivalent problem with four visible triangles. The conclusion is obvious: the small triangle is 1/4 of the large triangle.

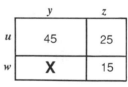

15. If A is $\Big|$, B is $\bigcirc$, C is $\quad$—$\quad$, D is $\bigcirc$, then $A * D$ is choice (a),

and $A * C$ is choice (d).

16. $\dfrac{u}{x} \cdot y = 45$, $\dfrac{u}{x} \cdot z = 25$, and $w \cdot z = 15$

Rectangle $X : w \cdot y = \dfrac{(w \cdot z)\,(u \cdot y)}{(u \cdot z)}$

$\qquad = \dfrac{15 \cdot 45}{25} = 27.$

	y	z
u	45	25
w	**X**	15

17. I recommend giving students a square or a rectangle to start with.

a)

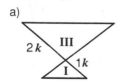

b)

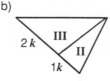

c)

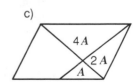

Similar triangles: sides, 2:1; areas, 4:1

Sides from part a): bases, 2:1; heights same; so areas are 2:1

Ratio of areas so far . . .

Region IV can be determined by observing IV + A = $4A$ + $2A$, IV = $5A$
I : II : III : IV = 1 : 2 : 4 : 5; since the area of the whole is 60, the areas are:
5, 10, 20, 25.
Extensions:

     Given lengths a and b, and $a \parallel b$

18. *Hint:* Graph and encase figures in a rectangle with sides parallel to the *x*-axis
and *y*-axis. Form rectangles and triangles and compute their sides from
coordinates.   *a*) 16   *b*) 30   *c*) Isosceles trapezoid: by use of
slopes and distance formula.   *d*) If coordinates are rational, encasement in
a rectangle ensures that all horizontal and vertical segments will be rational.
Thus, the areas of all rectangles and triangles will also be rational.

19. Distance formula: AB = 15 and AC = 25, so $AB : AC$
= 3:5, which divides BC in a ratio of 3:5. The *x* component is 24, so $3K + 5K$ = 24 and K = 3. Thus the
x coordinate of P is 9 ($3K$) to the right of -12 or x =
-3. The *y* component is 8; $3H + 5H$ = 8, or H = 1,
so the *y* coordinate of P is 3 ($3H$) above 2, or y = 5.
P is $(-3,5)$.

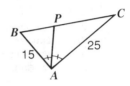

20. Area: $X \cdot Y$ = 22   Diagonal: $X^2 + Y^2$ = 100
$(X + Y)^2 = X^2 + Y^2 + 2(XY)$ = 100 + 2(22) = 144, so $X + Y$ = 12 and
perimeter = 24.

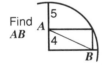

Find AB

21. I so enjoyed the famous rectangle-in-a-quarter-circle
problem at the left that I created problem 22 like it.
a) Draw diagonal SA; find PQ = 13.
b) Shaded = Whole – Squares = 91

22. Parallelogram: The segment joining the midpoints of two sides of a triangle is parallel to and
half the third side of the triangle. A pair of opposite sides of the quadrilateral formed are parallel and equal, so the figure is a parallelogram.

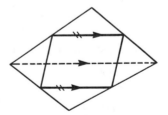

Answers to extensions: rhombus; rectangle;
rhombus; square; rectangle. (*Note:* We could
also ask, What must be true about the original
quadrilateral so the figure formed by joining consecutive midpoints will be a
rectangle? A square? A rhombus? Etc.?)

23.– 24. The maximum area for a given perimeter is the most regular figure (a
square). Similarly for minimum perimeter for given area. Many such maximum
and minimum problems can be handled by tables. Try for the most efficient

cylindrical can or other traditionally calculus-based problems. *Hint:* Encourage able students to write a computer program to obtain the raw data and then analyze the results.

25. *Method A.* $R^2 = r^2 + 25 \rightarrow R^2 - r^2 = 25$
    Area $= R^2\pi - r^2\pi = 25\pi$

Method B. It seems that the answer is independent of the size of the lighthouse or of the rug. Let the rug have radius 0, and so the stick is the diameter of the remaining circle. Thus the area is 25π.

26. *a)* $6 - x + 8 - x = 10$
$$2x = 4$$
$$x = 2$$

Notice "hinge effect."

$a + b = 10,$
$2a + 2b = 20$
4 left for $2x$

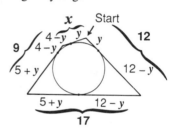

b) *Procedure:* "Walk around" the figure, labeling things as you go.

27. Walk around again.
$(4 - y) + y = x$
so $x = 4$

Notice: The sum of opposite sides is the same.

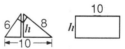

28. *Method A.* $x^2 + y^2 = 36$ *Method B.* Divide and conquer.

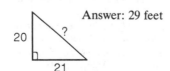

$(10 - x)^2 + y^2 = 64$ Right $\triangle$: $10 \cdot h = 6 \cdot 8$; $h = 4.8$
Solve for y and find area. $A = 24 + 10 (4.8) = 72$
(Very hard method.)

29. Definition of ellipse: Sum of $FA + GA$ = constant = major axis = 10. Thus, the sum of all six lengths is 30: $GA + GB + GC = 18$.

30. Unwrap the helix: Answer: 29 feet

31. *Solve:* $n(n - 3)/2 > 180 (n - 2) \rightarrow n^2 - 363n + 360 > 0$
Using the quadratic formula: $n < 2$ or $n \geq 362$ (hard algebra). A more creative

student wrote a short computer program to list *n,* number of diagonals, and sum of angles, and simply observed that the pattern switched for $n \geq 362$.

32. Will hit *P* again in next circuit because of the proportionality of sides when cut by a parallel to a side. Can trace through and verify.

33. See problem 7 and "fold" pyramid to form the same figure. Answer: 20

34. Draw an altitude to form two right triangles, and then draw a median to each hypotenuse (which is half the hypotenuse). The isosceles triangles can be rearranged.

35. Draw $\overline{OP}$ and double it to *Q.* Construct parallels to the sides of the angle from *Q* to form a parallelogram with diagonals $\overline{OP}$ and $\overline{XY}$.

36. Draw $\overline{BD}$ so that $\angle ABD \cong \angle BCA$ (*D* is between *A* and *C,* since $m\angle ABC > m\angle BCA$). $\triangle ABD$ is similar to $\triangle ACB \rightarrow x/4 = 4/6$; $x = 8/3$; $w/5 = 4/6$, so $w = 10/3 = y$, and $\angle DBC \cong \angle C \cong \angle ABD$, so $\angle ABC = 2 \angle BCA$.

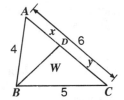

37. Segments form arithmetic progression. $7 : 7 + a : 7 + 2a : 18$. So $7 + 3a = 18$; $a = 11/3$. Thus, segments are 10⅔ and 14⅓.

38. Draw diagonal and divide each triangle 2:1 by a segment through a trisector of the base. These segments will trisect triangular areas.

BIBLIOGRAPHY

Artino, Ralph A., Anthony M. Gaglione, and Niel Shell. *The Contest Problem Book IV: Annual High School Mathematics Examinations 1973–1982.* Washington, D.C.: Mathematical Association of America, 1983.

Coxeter, H. S. M., and S. L. Greitzer. *Geometry Revisited.* Washington, D.C.: Mathematical Association of America, 1967.

Higgins, Agnes. *Geometry Problems.* Portland, Maine: J. Weston Walch, 1971.

National Council of Teachers of Mathematics. *Problem Solving in School Mathematics.* 1980 Yearbook of the NCTM. Reston, Va.: The Council, 1980.

Posamentier, Alfred, and Charles Salkind. *Challenging Problems in Geometry.* Vols. 1 and 2. New York: Macmillan, 1970.

Rhoad, Richard, George Milauskas, and Robert Whipple. *Geometry for Enjoyment and Challenge.* Evanston, Ill.: McDougal, Littell & Co., 1984.

Saint Mary's College. *Mathematics Contest Problems.* Palo Alto, Calif.: Creative Publications, 1972.

Salkind, Charles T. *Annual High School Mathematics Examinations,* Vol. 1, 1950–1960. Washington, D.C.: Mathematical Association of America, 1961.

_____. *The Contest Problem Book II: Annual High School Contests 1961–1965.* Washington, D.C.: Mathematical Association of America, 1966.

Salkind, Charles T., and James M. Earl. *The Contest Problem Book III: Annual High School Contests 1966–1972.* Washington, D.C.: Mathematical Association of America, 1973.

Stepelman, Jay. *Milestones in Geometry.* New York: Macmillan, 1970.

8

Logo Adds a New Dimension to Geometry Programs at the Secondary Level

Margaret J. Kenney

Geometry, as it is traditionally taught in the secondary school, is ripe for change. The time has come to reflect on and describe its evolution over the past two thousand years and to realize that it should also embody the technology of the present age. Geometry students should learn how geometric ideas and concepts apply to a wide range of human endeavor—in science, in art, and in the marketplace. Furthermore, students should experience geometry actively. One way to provide such experience is through the infusion of the computer into the curriculum. Logo, the programming language developed at Bolt, Beranek, and Newman and the Massachusetts Institute of Technology Artificial Intelligence Laboratory in the 1960s, is an excellent means of communicating with the computer.

Logo, a derivative of the list-processing language LISP, is for everyone. Originally, Logo evolved from the premise that programming can be an effective means of educating young children. However, it is also a highly sophisticated language whose potential has yet to be reached. Turtle graphics—the rendering of computer graphics through the movements of a triangular shape called a turtle—undoubtedly accounts for the success and appeal of the language with young learners. There is reason to believe this success can be matched with older students.

Logo is more than a programming language. It is a language that—

- promotes learning through discovery;
- develops problem-solving skills;
- supports the teaching of geometry.

Turtle graphics can be used effectively to enhance, and probe more deeply into, geometric topics on two levels. For those students whose experience with Logo is limited, teacher-prepared interactive activities can be woven into the curriculum. For the growing population of students who have had prior Logo exposure, more challenging work entailing the writing of procedures provides a pleasant change of pace in the typical geometry class routine. In either group, Logo is a tool—a tool that is intended to make geometry come alive for the student.

The following list of suggestions is a beginning. The list contains ideas for implementing Logo activities with students in low-level, standard, and advanced geometry classes. These examples are intended to serve as a catalyst for finding other ways to incorporate Logo with geometry.

Use Logo to see something new in a familiar problem or situation.

The use of Logo as an instructional tool in the secondary geometry classroom affords numerous opportunities to view again and then extend concepts and terms from informal geometry. That is, Logo provides an environment for the student to work on something familiar but from a different perspective. In the process, reinforcement occurs naturally, but often new relationships and insights also result. One significant instance of this occurs in the preliminary stages of using turtle graphics, namely, when the student attempts to make different polygons. To draw a triangle on paper with ruler and protractor requires a knowledge of the side lengths and angle measures of the triangle, that is, the angles interior to the figure. To draw a triangle on the screen using turtle graphics requires a knowledge of the side lengths and the angle measures exterior to the figure.

A triangle whose sides measure 87, 100, and 50 units with angles of 30, 60, and 90 degrees (see fig. 8.1) can be produced by the following sequence of steps: FORWARD (FD) 87 RIGHT (RT) 150 FD 100 RT 120 FD 50 RT 90. Thus, when students are using the commands FD, BACK (BK), RT, LEFT (LT), knowing the supplements and complements of angles is a requisite for producing arbitrary polygons.

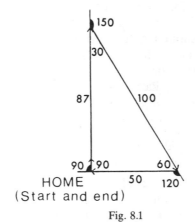

Fig. 8.1

The use of Logo in a familiar situation readily leads to new questions, many of which can be thoughtful and challenging. For example, take the well-known problem of finding the number of diagonals in a convex *n*-gon. Let this problem be the base for a number of Logo investigations. Here are some typical activities to try, arranged in order of difficulty.

Write a procedure to exhibit—

a) a square and its diagonals;

b) a regular pentagon and its diagonals;

c) a regular hexagon and its diagonals;

d) a regular *n*-gon and its diagonals.

Figure 8.2 shows the output of (*d*) when *n* = 6, 7, or 15.

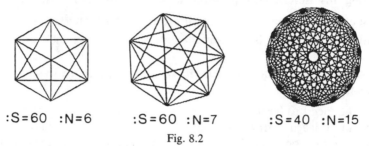

:S=60 :N=6   :S=60 :N=7   :S=40 :N=15

Fig. 8.2

The value of these activities is not just the attractive graphic display. What makes the exercise worthwhile is the opportunity to apply previously learned content, the effective use of problem-solving skills, and the realization that the problem can be solved by a variety of approaches.

The specific procedure chosen for (*d*) relies on the distance formula and on the fact that the same number of diagonals emanate from each vertex of the polygon:

Main Driver Procedure

(:S represents side length, :N represents number of sides)

```
TO DIAGONALS.NGON :S :N
HT
PU
SETXY − 100 0
PD
FULLSCREEN
REPEAT :N [MAKE "K :N − 2 MAKE "X XCOR MAKE "Y YCOR DIAGONALS :S :N 0
    FD :S RT 360/:N FD :S RT 360/:N]
END
```

Subprocedure

(:D is a counter)

```
TO DIAGONALS :S :N :D
IF :D = :N − 1 THEN STOP
FD :S
MAKE "XJ XCOR
MAKE "YK YCOR
```

```
SETHEADING TOWARDS :X :Y
FD SQRT ( ( :XJ − :X) * ( :XJ − :X) + ( :YK − :Y) * ( :YK − :Y) )
BK SQRT ( ( :XJ − :X) * ( :XJ − :X) + ( :YK − :Y) * ( :YK − :Y) )
LT :K * ( ( 180 − 360 / :N) / ( :N − 2) )
MAKE "K :K − 1
DIAGONALS :S :N :D + 1
END
```

The student should be aware that the procedure in *(d)* merely illustrates the solution to the problem. However, data accumulated from systematically executing this procedure as well as from the procedural development itself are useful for deriving a formula for the number of diagonals in a convex *n*-gon.

During this exercise new questions arise for additional study, such as:

- In a convex *n*-gon, how many different diagonal lengths are there?

- In a convex *n*-gon, how many diagonals are there of each length?

Use Logo to encourage the student to do one problem in several different ways.

Students need to be directed to solve a problem in more than one way. Much can be learned from the analysis of different approaches to doing a problem. In fact, occasional class discussions focused on the variety of ways of solving a problem are a particularly effective format for the review and integration of content. The Logo language is ideally structured for encouraging students to seek more than one way of getting a solution. For example, each of the following procedures generates a right triangle with legs of *a* units and *b* units.

```
TO RIGHTRI1 :A :B                TO RIGHTRI2 :A :B
  SETXY :A 0                        RT 90
  SETXY :A :B                       FD :A
  HOME                              LT 90
  HT                                FD :B
END                                HOME
                                   HT
                                 END
```

Some Logo primitives inhibit the problem solver's need to think mathematically. Thus, it becomes more of a challenge to restrict the use of certain commands. Consider right triangle procedures subject to these conditions:

1. Suppose the command HOME is disallowed. One workable alternative is as follows:

```
TO RIGHTRI3 :A :B
SETXY :A :B
SETXY :A 0
SETXY 0 0
HT
END
```

2. Suppose the commands SETXY and HOME are disallowed. This alternative calls for the use of the Pythagorean theorem:

```
TO RIGHTRI4 :A :B
RT 90
FD :A
LT 90
FD :B
SETH TOWARDS 0 0
FD SQRT ( :A * :A + :B * :B )
HT
END
```

3. Suppose the commands SETXY, HOME, and SETHEADING TO-WARDS are disallowed. This alternative makes use of interior/exterior angle relations in a triangle, inverse trigonometric functions, and the Pythagorean theorem:

```
TO RIGHTRI5 :A :B
RT 90
FD :A
LT 90
FD :B
LT 90 + ATAN :B :A
FD SQRT ( :A * :A + :B * :B )
HT
END
```

Use Logo to visualize problems with a construction emphasis.

Every teacher of geometry experiences frustration at being unable to explain certain concepts adequately because of a lack of good sketches. Logo can readily be used for instructional support by providing appropriate graphics in many circumstances. Indeed, some displays are so effective, they belong in the category of "proofs without words." In figure 8.3, a circle is clearly observed to be the limiting case of a sequence of n-gons, with n increasing.

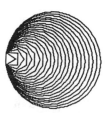

Fig. 8.3

The procedure and subprocedures that produce this figure are as follows:

Main Driver Procedure

```
TO CIRCLELIMIT
  MAKE "N 3
  LIMCIRCLE
END
```

Subprocedures

```
TO LIMCIRCLE
  IF :N > 25 STOP
  POLYSTART
  MAKE "N :N + 1
  LIMCIRCLE
  FULLSCREEN
END

TO POLYSTART
  PU SETXY - 50 0 PD
  REPEAT :N [FD 20 RT 360/:N]
  HT
END
```

Use Logo to test conjectures.

Conjectures are an integral part of the development of mathematics, and they play a significant role in the history of mathematics. Many conjectures framed centuries ago still stand as unresolved challenges for the present-day mathematical community. Students should be exposed to some of these well-known conjectures and also be encouraged to formulate conjectures of their own in the various branches of mathematics. A Logo environment is a fertile testing ground for working with geometric conjectures.

Consider this example:

> Given an isosceles right triangle with legs of 1 unit, assume that one leg lies along the positive x-axis, as in figure 8.4a. Erect a second right triangle on top of the first with legs $\sqrt{2}$ units and 1 unit, as in figure 8.4b. Continue the process of building a new right triangle on top of the preceding one.

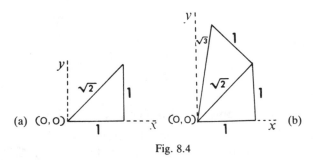

Fig. 8.4

CONJECTURE. *Twelve or fewer right triangles are required before the hypotenuse of the last triangle crosses over the positive x-axis.*

This problem is based on one attributed to Theodorus, who was born about 465 B.C. and became the mathematical tutor of Plato. The construc-

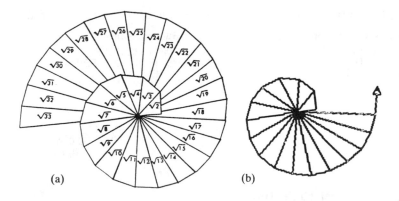

Fig. 8.5

tion in figure 8.5a shows rational and irrational lengths $\sqrt{n}$, n a positive integer < 34, in a spiral pattern. Figure 8.5b gives the Logo version of the problem. Each diagram clearly indicates the conjecture is false. To obtain an accurate sketch similar to figure 8.5a, the student must exercise care in drawing each right triangle. To obtain the output shown in figure 8.5b, a Logo procedure can be developed using the powerful and invaluable idea of recursion. Try the following:

Main Driver Procedure

```
TO TWHEEL
  MAKE "S 1
  RT 90
  FD 20
  LT 90
  FD 20
  BUILD :S
END
```

Subprocedure

```
TO BUILD :S
  IF :S > 16 THEN STOP
  SETH TOWARDS 0 0
  FD 20 * SQRT ( :S + 1 )
  BK 20 * SQRT ( :S + 1 )
  RT 90
  FD 20
  BUILD :S + 1
END
```

Use Logo to provide practice in looking for patterns that might otherwise be inaccessible.

Logo is a helpful tool in the investigation of many problems that can be handled only in limited fashion with paper and pencil. One area in particular that can be expanded considerably is spirolaterals. This topic, an interesting blend of number theory and geometry, is rich with unexplored activities.

Spirolaterals, or spirograms, are a series of connected line segments formed by a given number sequence, turning angle, and direction. Graph paper and square or isometric dot paper make a study of spirolaterals possible for turning angles of 45, 60, 90, and 120 degrees and some choices of number sequences. In general, though, paper and pencil inhibits discovery because results come slowly, even for the previously mentioned angles. Logo procedures, many of which involve recursion, provide data readily for a variety of spirolateral problems. They also stimulate the student to try more complicated extensions.

For example, use the procedure TEST.FIVE—

1. to determine what choice of input for :A :B :C :D :E produces a design that is a square dissection;

2. given that a square dissection pattern is found, to find the length of a side of the square in terms of :A :B :C :D :E.

```
TO TEST.FIVE :A :B :C :D :E
  RT 90
  REPEAT 4 [FD :A RT 90 FD :B RT 90 FD :C RT 90 FD :D RT 90 FD :E RT 90]
  PU
  HOME
  PD
  HT
END
```

Figure 8.6a shows a square design with side length of 60 units, resulting from an input of 40, 10, 40, 50, 20 units. Figure 8.6b shows a closed but nonsquare design whose input is 10, 20, 30, 40, 50 units.

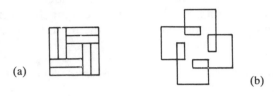

(a)

(b)

Fig. 8.6

Use Logo in making generalizations or drawing conclusions.

Logo also serves as a viable means of making generalizations and drawing conclusions in those geometry courses that do not stress proof. Courses that de-emphasize proof do expect students to have a reasonable awareness and understanding of the facts found in the statements of theorems. Some very effective discovery activities can be formulated by reworking the statements of theorems. It is possible to structure the activities so that even the student who knows very little Logo can participate fully.

Activity: To discover that the ratio of the diagonal of a square to one of its sides is constant.

Give the student—

1. the procedure SQ :S

```
TO SQ :S
  REPEAT 4 [FD :S RT 90]
  RT 45
  HT
END
```

2. a table to complete like the one in figure 8.7.

:D in the table is to be estimated for a given :S using the procedure SQ :S. This is accomplished by making trial-and-error moves from the lower left to the upper right corner of the square that appears on the screen. Sufficient care must be taken to get the line segment to meet the upper right vertex without extending beyond it. The student will not learn that $\sqrt{2}$ is the constant from this exercise, but those who estimate carefully enough will be able to deduce a constant value of 1.41.

Similarly, students can find that the ratio of the circumference of a circle to one of its diameters is constant, and that the hypotenuse and one leg in a 30-60-90-degree triangle have a very apparent connection. Actually, the traditional geometry text is replete with theorems that make excellent activities. A good challenge for those students who know some theorems and

Complete the table. Use Logo to guess the length :D. Use your calculator to find the ratio :D/:S correct to hundredths.

:S	:D	Ratio :D/:S
40		
50		
60		
70		
80		
90		
100		

Study the results above carefully. Look for patterns or relations. Write a sentence that describes your findings.

Fig. 8.7

Logo is to have them prepare some of these discovery activities for less able groups of students to use.

Use Logo to apply motion geometry concepts.

A study of motion geometry is greatly enhanced by the use of Logo. Here we consider one aspect only—the application of motion geometry concepts to tessellations.

Tessellations, or tiling patterns, are popular and surround us in a variety of modes, such as floor coverings, ceilings, wallpaper, fences, and cloth, to name a few. But tessellations require time and patience to design and complete. Square and isometric dot paper are helpful aids for getting started. However, they often dominate the finished product or limit the student's choice of shapes. A possible alternative is to have the student teach the computer to tessellate and let the printer do the drawing. Such an activity effectively combines concepts of motion geometry with strategies of problem solving in a meaningful application for the student. Rotations, reflections, and translations, the backbone of motion geometry, are used automatically in the course of producing the tessellation. Logo's procedural development is ideally suited to this problem. The student does not need to have an advanced knowledge of Logo to be able to produce interesting results. Once an overall plan for developing a particular tessellation takes root, then many other pleasing designs can be generated with slight modifications of the original plan.

To get started, provide students with a completed tessellation to study. Spend some time examining it and discussing how it might be drawn. For example, if the tessellation is the familiar tiling of squares, the student should observe that it can be produced by—

a) drawing all the horizontal line segments first, and then drawing all the vertical line segments;

b) drawing the top row of squares, then the second row, third row, and so on;

c) drawing the leftmost column of squares, then the second column, third column, and so on.

Be sure all the students are aware of the limitations of a method like *(a)*. Point out that several other methods may be equally appropriate, like *(b)* and *(c)*. Be sure the students also realize that often line segments can be drawn and redrawn several times over in the process, as in *(b)* and *(c)*, and that this action does not change the finished product.

For some students, a preliminary discussion will be sufficient preparation for the task, and they will be ready to try their own procedure. Others will benefit from a more detailed model to imitate, such as the procedural plan (see fig. 8.9) for the patterned square tessellation in figure 8.8.

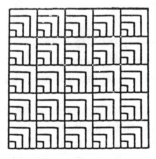

Fig. 8.8

Once the student has mastered the format of the procedural plan, then more complex patterned squares can be used to generate tessellation designs. Two of these appear in figures 8.10 and 8.11. This model can also be extended to create Escher-like tessellations such as the cats shown in two stages in figure 8.12. Figure 8.12a demonstrates that the first stage is actually a patterned square; in figure 8.12b two sides of each square have been removed so the cats will dominate the tessellation.

Procedural Plan

1. Make the basic shape, here a patterned square. Use
   TO SQUARE :S
      REPEAT 4 [FD :S RT 90]
   END.

```
TO FOURSQUARE
   SQUARE 10
   SQUARE 20
   SQUARE 30
   SQUARE 40
END
```

2. Move the turtle so that two adjacent patterned squares can be drawn with a common side.

```
TO MOVE.1
   PU
   RT 90
   FD 40
   LT 90
   PD
END
```

3. Make a row of five patterned squares.

```
TO ROW.SQUARE
   REPEAT 5 [FOURSQUARE
             MOVE.1]
END
```

4. Identify the starting point for the tessellation.

```
TO POSITION
   PU
   SETXY - 100 60
   PD
END
```

5. Be able to move from the end of one row to the beginning of another.

```
TO MOVE.2
   LT 90
   FD 200
   RT 90
   PU
   BK 40
   PD
END
```

6. Produce the five-row tiling, using the **main driver procedure.→**

```
TO FOUR.SQUARE.TESSELLATION
   HT
   POSITION
   REPEAT 5 [ROW.SQUARE
             MOVE.2]
   FULLSCREEN
END
```

Fig. 8.9

Fig. 8.10   Fig. 8.11

The driver procedure and its component subprocedures are as follows:
CATS.TESSELLATION
| | POSITION (sets starting point of the tessellation)
|ROW.CAT (forms a row of face up/down cats)—MOVE2 (from one
| row to the next)
|CAT— MOVE1 (aligns cat pairs side by side)
(↑ the longest procedure that makes the cat pair outline
within a square.)

The possibilities for tessellations with Logo are restricted only by the doer's
creativity and persistence.

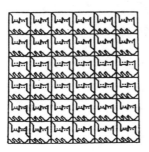

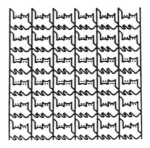

Fig 8.12

Use Logo as a designer's tool.

Applications of mathematics to art cover a wide spectrum of choices,
ranging from those that can be carried out successfully by all students to
others that are suitable only for students with a special artistic flair. Applying
mathematics to art with Logo allows many students who are not talented

with paper and pencil to participate. Both applications described below depend on mathematics to achieve the finished product.

Logos with Logo. Students learn to sharpen their awareness skills when they are asked to duplicate familiar logos or trademarks of companies, associations, and groups. Further, they often must accommodate or refine the logo to fit the limitations imposed by turtle graphics. Developing a collection of logos makes a good class project that at the same time evolves into an attractive bulletin board display (see fig. 8.13). A viable extension of the project is to encourage students to create their own logos.

Fig. 8.13

Quilting designs. Quilts abound with intricate geometric patterns. A closer look reveals that many are a consequence of implementing rotations, reflections, and translations. A variety of quilting designs can be created with the use of Logo and the idea of a Latin square. A Latin square is an $n \times n$ square array in which n distinct elements are arranged so that each element occurs exactly once in each row and column.

$$
\begin{array}{cccc}
1 & 2 & 3 & 4 \\
4 & 3 & 2 & 1 \\
2 & 4 & 1 & 3 \\
3 & 1 & 4 & 2
\end{array}
$$

For example, this array is a 4×4 Latin square. Four different patterned squares can be placed together to form Latin squares as in figure 8.14. The 8×8 quilting designs in figure 8.15 are formed by rotating each 4×4 array about its lower-right vertex. Clearly, many other quilting patterns can be generated from the Latin squares in figure 8.14 or from different Latin square formations of the same four patterned squares. This is just the beginning of the quilting exploration. Interested students can produce some really spectacular designs.

Fig. 8.14

Fig. 8.15

Use Logo to build models for geometric series.

Geometric series are more appealing when they are associated with a specific situation or model. One familiar and fascinating example of this is Helge von Koch's snowflake curve. The first stage of this curve is an equilateral triangle. Figure 8.16 illustrates the first four stages with sizes modified so that all can be shown together. The snowflake curve is remarkable because it encloses a finite area and has an infinite perimeter. In spite of the admonition "Divergent series are the invention of the devil, and it is a shame to base on them any demonstration whatever," attributed to the mathematician Abel, curves that involve the infinite do generate considerable interest.

Fig. 8.16

Logo, with its recursive nature, ideally permits similar studies of many regular polygons and other shapes. Investigate the iterative yield when straight line segments are replaced with one of these:

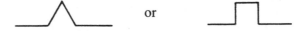

For example, follow the sequence of shapes begun in figure 8.17. These are an example of a fractal, the name given by Benoit Mandelbrot in 1975 to a class of highly irregular shapes such as the von Koch snowflake. The procedure that yields the fractal based on a square is as follows:

Main Driver Procedure (:X represents side length, :Y represents Stage − 1)

```
TO SQ.FRACTAL.S :X :Y
   CENTER2 :X
   REPEAT 4 [FRACTAL :X :Y RT 90]
END
```

Subprocedures

```
TO CENTER2 :X
   CS
   PU
   HT
   SETXY − :X / 2 ( − :X / 2)
   PD
   FULLSCREEN
END
```

```
TO FRACTAL :X :Y
   IF :Y = 0 THEN FD :X STOP
   FRACTAL :X / 3 :Y − 1
   LT 90
   FRACTAL :X / 3 :Y − 1
   RT 90
   FRACTAL :X / 3 :Y − 1
   RT 90
   FRACTAL :X / 3 :Y − 1
   LT 90
   FRACTAL :X / 3 :Y − 1
END
```

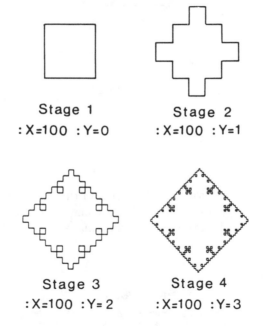

Stage 1
: X =100 : Y=0

Stage 2
: X=100 : Y=1

Stage 3
: X=100 : Y= 2

Stage 4
: X=100 : Y= 3

Fig. 8.17

If the side of the square in stage 1 is a units, then the geometric series representing the perimeter of the fractaled shape is

$$4a + \frac{8a}{3} \left(1 + \frac{5}{3} + \left(\frac{5}{3}\right)^2 + \ldots + \left(\frac{5}{3}\right)^{n-1} + \ldots\right).$$

Since $5/3 > 1$, the series diverges and the perimeter is infinite. The geometric series corresponding to the area of the fractaled shape is

$$a^2 + \frac{4}{9} a^2 \left(1 + \frac{5}{9} + \left(\frac{5}{9}\right)^2 + \ldots + \left(\frac{5}{9}\right)^{n-1} + \ldots\right).$$

Since $5/9 < 1$, the series converges and the area is finite. Although the fractaled square has an infinite perimeter, it has a finite area of $2a^2$. Logo provides clear evidence that the area of the fractaled square is twice that of the original square, as shown by the superposition of stage 1 on stage 4 in figure 8.18.

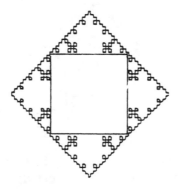

Fig. 8.18

The reader is invited to probe further and discover still other ways in which Logo can support and enrich the teaching of geometry. Discovery— that is, seeing what all have seen and thinking what none have thought— awaits you.

BIBLIOGRAPHY

Bezuszka, Stanley J., Margaret J. Kenney, and Linda Silvey. *Designs from Mathematical Patterns*. Palo Alto, Calif.: Creative Publications, 1978.

———. *Tessellations: The Geometry of Patterns*. Palo Alto, Calif.: Creative Publications, 1977.

Billstein, Rick, Shlomo Libeskind, and Johnny W. Lott. *Logo, MIT Logo for the Apple*. Menlo Park, Calif.: Benjamin Cummings Publishing Co., 1985.

Mandelbrot, Benoit B. *Fractals: Form, Chance, and Dimension*. San Francisco, Calif.: W. H. Freeman & Co., Publishers, 1977.

Papert, Seymour. *Mindstorms: Children, Computers, and Powerful Ideas*. New York: Basic Books, 1980.

9

Some Modern Uses of Geometry

Donald W. Crowe
Thomas M. Thompson

FOR more than a century the readers of high school geometry texts might have thought that the main use of geometry was as a vehicle for studying logic and the nature of proof. This emphasis in geometry texts had a natural historical background in two related ways.

First, the "axiomatic method" was invented, or at least first presented in its full-blown form, by Euclid in *The Elements*. Since *The Elements* also contains a systematic collection of basic geometric facts, its geometric part came to be accepted as the standard secondary school geometry text in many parts of the world. Thus the axiomatic method—which by the late nineteenth century came to be seen as the underlying logical tool for *all* mathematics—was associated especially with geometry in the minds of those who went no further into mathematics than what they learned in high school.

Second, the development of non-Euclidean geometries, beginning with Bolyai and Lobachevski in the early nineteenth century, served to further emphasize the connection between philosophy and geometry. The background for this was that the influential philosopher Immanuel Kant was thought to have taught that the human mind was structured in such a way that no geometry other than Euclid's would fit into it. Bolyai and Lobachevski demonstrated that their non-Euclidean "hyperbolic geometry" could be comprehended and developed equally well by the human mind. This became the dominant point of view when it was further realized that ordinary spherical geometry, when slightly modified to become "elliptic geometry," fit naturally into the same framework. All three geometries were based on essentially the same axiom system, with one of the axioms, the "parallel axiom," being suitably modified to yield each of the three. Euclidean geometry occupied a delicate middle position ("unique parallels") between hyperbolic ("many parallels") and elliptic (no "parallels") geometries.

The lingering feeling among mathematicians that there might still be some inherent contradiction in the new hyperbolic geometry vanished when it was proved later in the century that hyperbolic and Euclidean geometry were equally consistent. That is, a model of each could be constructed within the other; so any inconsistency in one would immediately be revealed as an inconsistency in the other. From this time on, mathematicians, and the textbook writers influenced by them, used this and other examples from geometry to illustrate a variety of logical ideas associated with the axiomatic method.

However, the ordinary citizen knew little of this. In the nineteenth century few people went to high school or had contact with Euclid's geometry. Even the elite, the artisans with their special knowledge, knew little of the "logic" of Euclid. Carpenters continued to build by rules of thumb—sometimes "correct" from the Euclidean point of view, sometimes not. Wine barrels that held more or less the amount they were believed to hold continued to be built. Today there are new artisans and technicians who may be designing computer chips or robots, studying the structure of viruses, designing and building geodesic domes, or planning experiments to test the efficacy of new fertilizers on corn fields.

What do these modern artisans know, or need to know, of the logical foundations of Euclidean geometry? Very little, in most instances, and yet they used geometry in other ways more essential to their work. It is our purpose here to indicate some of those ways in which geometry and geometric modes of thought are used—or could be used—by these modern artisans.

Finite Geometries and the Design of Experiments

In the late nineteenth century, a few simple finite geometries were invented for testing or demonstrating the "completeness," "consistency," or "independence" of various modifications of Euclid's axioms. For example, suppose we want to show that the primitive notions of point and line, together with the following three axioms A1, A2, and A3, are consistent.

A1: *Every pair of points lies on one and only one line.*

A2: (Euclidean parallel axiom) *For each line and each point not on the line, there is exactly one line containing the point and having no point in common with the line.*

A3: *There are (at least) four points, no three of which are on the same line.*

To show the consistency of these axioms, it is enough to produce a "model." Such a model, with exactly four points, *A, B, C, D,* and exactly six lines, *AB, AC, AD, BC, BD, CD,* with two points lying on each line and three lines through each point, is shown in figure 9.1.

Any set of "points" and "lines" satisfying A1–A3 is called an *affine plane* and is said to have *order n* if there are exactly n points on each line and exactly $n + 1$ lines containing each point. There are infinitely many values of n (but not *all n*) for which such a model can be produced. Each model

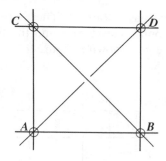

Fig. 9.1

should be thought of as being a rudimentary "Euclidean plane," without any congruence, "betweenness," or continuity. Modern mathematicians are not as interested in the "logical" problems associated with these geometries as they are in such questions as these:

Question 1. *(a)* For what values of n is there such a geometry? *(b)* In particular, is there an affine plane when $n = 10$?

Question 2. If there is an affine plane for a given order n, can we describe all the different ones?

The value $n = 10$ is the least one for which question 1(a) is still unsolved. The least n for which question 2 is unsolved is 9. There are four known planes of order 9, but nobody knows whether there are others. For at least twenty-five years many of us have imagined that the answer to question 1(b) is just around the corner. But it remains unsolved, though much is now known about what such a geometry would look like, *if there is one.*

What shifted modern interest from the logic and philosophy of finite geometries to more prosaic questions like 1 and 2? The answer is that finite geometries have applications in *statistics,* to the "design of experiments," which involves the construction of so-called orthogonal Latin squares. We shall illustrate this with a simple example, letting $n = 5$. (See Beck, Bleicher, and Crowe [1969, chap. 4] for more details.)

For this case ($n = 5$) it is very easy to find an affine plane with exactly five points per line. (Incidentally, there is nothing special about five; an analogous construction is easily carried out when n is any prime number.) All that is needed is to use the five "integers modulo 5," 0, 1, 2, 3, 4, as if they were ordinary coordinates, to construct a coordinate plane having twenty-five *points.* The points are the twenty-five pairs (x,y) where each of x and y ranges over the five symbols 0, 1, 2, 3, 4. See figure 9.2, on which some of the points have been labeled. A typical *line* is $2x + 3y = 1$ (in general, $ax + by = c$ for any choice of a, b, c from the integers modulo 5). This line contains the five points (0,2), (1,3), (2,4), (3,0), (4,1) (all marked by boxes in fig. 9.2) and has "slope" 1. There are thirty lines altogether, including the five horizontal lines $y = c$ and the five vertical lines $x = c$ that form the coordinate grid shown in figure 9.2.

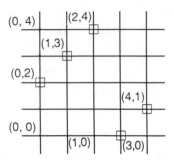

Fig. 9.2

Now suppose all the points on line $2x + 3y = 1$ are labeled by the letter a. (See fig. 9.3a.) Then pick some line parallel to it (e.g., $2x + 3y = 4$) and label all its points b. Finally, label the points of the other parallels c, d, e, respectively. This gives us a picture like figure 9.3b.

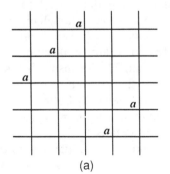 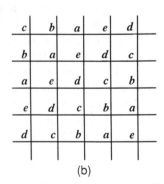

(a)   (b)

Fig. 9.3

This 5×5 array is an example of a *Latin square*. That is, each row and each column contains each of the five letters a, b, c, d, e exactly once. Picking any other parallel class of lines, such as lines of slope 2, yields a second Latin square, perhaps as in figure 9.4.

C	A	D	B	E
B	E	C	A	D
A	D	B	E	C
E	C	A	D	B
D	B	E	C	A

Fig. 9.4

These two Latin squares are *orthogonal* to each other in the sense that when they are juxtaposed, each of the twenty-five pairs aA, aB, . . ., eE appears exactly once in the resulting 5×5 array (fig. 9.5).

cC	bA	aD	eB	dE
bB	aE	eC	dA	cD
aA	eD	dB	cE	bC
eE	dC	cA	bD	aB
dD	cB	bE	aC	eA

Fig. 9.5

It is just this kind of array that statisticians need for the "design of experiments" of the following type. Suppose each of five kinds of fertilizer (a, b, c, d, e) is to be tested in conjunction with five dosages (A, B, C, D, E) of water in a field of corn. To randomize the experiment completely (e.g., to minimize the influence of shade trees or the residue from a previous experiment along one row of the field), it is desirable to arrange the twenty-five combinations so that all five different fertilizers and all five dosages of water are used in each row (and "column"). We see that the two juxtaposed squares of figure 9.5 accomplish just that, and the geometry has told the statistician how to do it!

The reader may think this is easy to do *without* the geometry, and for this case, $n = 5$, it is—and can be done on a napkin while having coffee. But it will be instructive for you to try $n = 6$. As you will find after the use of many napkins, *it is impossible to find two such orthogonal 6 × 6 squares.*

This means, in particular, that there is no affine plane of order 6. Why? Because if there were, then it would be possible to use two parallel classes of lines in that plane to construct two orthogonal 6 × 6 squares, very much as in our example for $n = 5$. It *is* possible to construct two orthogonal 4 × 4 squares, although it is not as easy as 5 × 5.

Whenever n is of the form p^t (p a prime number, t a positive integer), there is a finite arithmetic ("finite field") having n elements that can be used as coordinates to produce an affine plane of order n and hence two (in fact, $n - 1$) orthogonal Latin squares. More than two hundred years ago, Euler knew that there could not be two orthogonal 6 × 6 squares, and he thought that there could never be two orthogonal $(4k + 2) \times (4k + 2)$ squares for any k. It was a dramatic episode in the history of mathematics when this long-standing conjecture was proved *false* in 1958 by the mathematicians Bose, Parker, and Shrikhande. The two 10 × 10 squares that they produced are shown in color on the cover of *Scientific American* (Gardner 1959), which should be read for the whole fascinating story.

Notice, however, that the production of the two 10 × 10 squares does not yet answer question 1(b). Finding a whole affine plane of order 10 would involve the construction of *nine* mutually orthogonal 10 × 10 squares. The general question of whether there is any affine plane whose order is *not* of the form p^t remains unanswered to this day.

Transformation Geometry and Archaeology

One of the "modern" themes and approaches to geometry is transformation geometry, a way of looking at geometry that is more global than local. Instead of looking at individual triangles, circles, and polyhedra as Euclid did, transformation geometry concentrates on translations, rotations, and reflections—in short, the *symmetries* (i.e., isometries, rigid motions) that move these figures around.

We shall illustrate the value of this alternative approach to geometry by means of recent developments in archaeology. At certain archaeological sites, decorated pots (or broken pieces of such pottery) have been found, and traditionally a rudimentary sort of geometry has been used to study them. Frequently the design "motifs" are classified as triangles, circles, and the like. However, such motifs are so common that they often do not provide, by themselves, sufficiently fine distinctions to be a satisfactory classificatory tool. If, as often happens, the motifs are arranged in repeated patterns (bands, or allover patterns), then the whole pattern can be analyzed according to its symmetries. This symmetry classification then becomes an important supplement to the crude classification by individual motifs.

We shall describe how this symmetry classification works by considering the special instance of one-dimensional (band) patterns. Two such patterns are said to be different if the rigid motions that move each pattern *onto itself* are different. For example, consider the following infinite bands:

1. - - - - V V V V - - - -
2. - - - - N N N N - - - -

They are different because there are reflections (in vertical lines between the Vs) that transform pattern 1 onto itself, but no such reflections that transform pattern 2 onto itself. However, from this point of view, the patterns

1'. - - - - W W W W - - - -

and

2'. - - - - S S S S - - - -

are the same as (1) and (2), respectively. The reason is that both (1) and (1') admit translations and vertical reflections (and no other symmetries), whereas both (2) and (2') admit translations and half-turns (and no other symmetries).

A complete analysis shows that there are exactly seven of these repeated band patterns. Examples of the seven, copied from the pottery of San Ildefonso Pueblo in New Mexico, are shown in figure 9.6 (Chapman 1970). They are labeled there with a convenient notion that is determined by the symmetries of the pattern as follows:

The first symbol is **m** (for mirror) if there is a reflection in a vertical line; otherwise it is **1**. The second symbol is **m** if there is a reflection in a horizontal line, **g** if there is a glide reflection but no horizontal reflection, **2** if there is a half-turn but no horizontal reflection or glide reflection, and **1** otherwise. (A *glide reflection* is a translation along some line followed by a reflection in that line; it is a theorem of elementary transformation geometry that every rigid motion is either a reflection, translation, rotation, or glide reflection.)

It is also quite common that patterns (such as an infinite chessboard) have two colors that are *equivalent* in the sense that some motions take the pattern onto itself *provided that the colors are reversed* at the same time. If we take into account such "dichromatic" one-dimensional patterns, there are a further seventeen bands, which help us extend our cataloguing possibilities. Examples of these seventeen are shown in figure 9.7. Except for the three labeled **mg/1g, 1m/11,** and **mm/12,** they were all copied from the pottery of San Ildefonso. The other three were invented to complete the list, in San Ildefonso style.

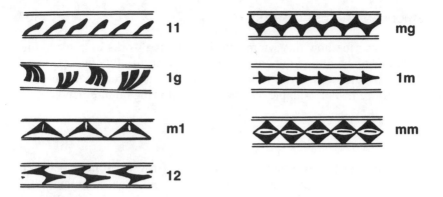

Fig. 9.6. Examples of all seven possible monochromatic band patterns, from the pottery of San Ildefonso Pueblo.

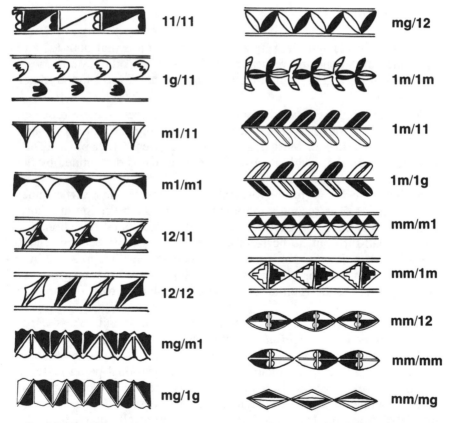

Fig. 9.7. Examples of all seventeen possible dichromatic band patterns. The symbol to the left of the "/" is the symmetry type of the corresponding monochromatic pattern. The symbol to the right is the symmetry type of the pattern consisting of either of the colors alone.

We conclude this section with a brief mention of two minor successes of this method of pattern analysis.

A 1970 British Museum exhibition ("Divine Kingship in Africa") contained a large display of art objects brought back from an 1897 British raid on the old Nigerian city of Benin. After that raid, illustrations of these spectacular objects, especially the bronze work, were widely reproduced. It was fairly well established that only thirteen of the seventeen mathematically possible two-dimensional patterns were to be found in the art of Benin (Crowe 1975). However, in this exhibit there was a mask whose pattern was very clearly that of one of the four missing types. On closer investigation it was found that this mask did not, in fact, originally come from Benin. It had found its way to Benin from a nearby community and had been indiscriminately taken and sent back to England by the British raiding party. (For an elementary discussion of other patterns in African art, see Zaslavsky [1973, chap. 14].)

A second, more interesting and complex example, is that of Chaco Canyon and sites near Chaco Canyon in northwestern New Mexico and southwestern Colorado. The following account is a very abbreviated version of a report by Washburn and Matson (1985).

The proportions of designs of the various symmetry types were calculated from field reports of twenty-five sites in the vicinity of Chaco Canyon. Then a "map" showing the probable location of each of the twenty-five sites was constructed by a statistical "multidimensional scaling" process based on the "distance hypothesis" that nearby populations will produce designs with greater similarity of symmetry structure than populations living farther apart. If the distance hypothesis was valid, then the resulting map would be in substantial agreement with the actual geographical map of these locations. The statistically generated map *was* in good agreement with the geographical map. This supports the "distance hypothesis" and suggests its usefulness in other similar situations.

The few sites that did not appear in their correct geographical locations also provide useful directions for further archaeological research. In some instances it had already been suspected that the sites traded more actively with other regions than with Chaco Canyon. In others, new research is needed to suggest what other factors (e.g., the difficulty of access, social reasons for the development of other symmetry patterns, etc.) are involved.

Geometry and Groups

The regular plane figures provide a simple yet elegant link from geometry to group theory. We shall illustrate this with a square. Rotating the square one quarter-turn counterclockwise has the effect of sending vertex 1 to vertex 2, 2 to 3, 3 to 4, and finally, 4 to 1. We denote this rotation by the letter f. Two such quarter-turn rotations, one following the other, shall be denoted by $f \cdot f$ or f^2. This operation should be viewed as a composition of functions (and not just multiplication). Both f and f^2 are shown in figure 9.8.

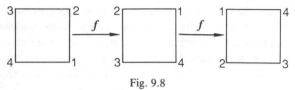

Fig. 9.8

Other rotations of the square include f^3 and $e = f^4$ (this last one, in effect, does nothing at all). There are four other *isometries,* or distance-preserving motions, of the square into itself. For example, let g fix vertices 1 and 3 and interchange vertices 2 and 4. This is a reflection about the axis passing through vertices 1 and 3. Further compositions of f and g yield a total of eight distinct isometries of the square. This set of eight elements, together with the binary operation "composition," forms the non-Abelian *dihedral* group of order 8:

$$D_4 = \{e, f, f^2, f^3, g, fg, f^2g, f^3g\}$$

(Recall that a group is a nonempty set together with an associative binary operation that has an identity and in which every element has an inverse. The integers form a commutative (Abelian) group under addition.)

Other such geometric shapes yield similar *symmetry* groups. In particular, any regular n-gon gives rise to the group D_n. In other dimensions the results are even more interesting. For example, in three dimensions the regular tetrahedron gives rise to the *symmetric* group on four elements, S_4, whereas the regular icosahedron (twenty sides) yields the symmetric group on five elements, S_5. Of particular interest are those figures that yield large simple groups (groups having no nontrivial normal[1] subgroups) as either the symmetry group itself or as some subgroup or factor group of the symmetry group. It has been known for many years (Thompson 1983, pp. 100–101) that the simple groups are in some sense the "building blocks" of all finite groups. Thus, a knowledge of all simple groups would be a giant step in the direction of the classification of all groups. For example, although the symmetry group of the icosahedron (having order 120) is not simple, its subgroup of order 60 is.

Although the study of simple groups had begun much earlier, the subject lay dormant until the 1950s. By 1965 Zvonimir Janko discovered a new simple group, and the race was on, with new ones being discovered in quick

---

1. A subgroup N of a group G is normal if whenever n is in N and g is in G then $g^{-1}ng$ is also in N.

succession until there were a total of twenty-six. Three of these simple groups were discovered by John Conway in 1968, all derived from a single symmetry group of a rather complex figure in E^{24} (twenty-four-dimensional space). The figure, which we shall call P for "polytope" (the generic n-dimensional analog of polygon or polyhedron), has 196 560 vertices, each the same distance from the center. The symmetry group of P is extremely large, having an order of

$$2^{22} \cdot 3^9 \cdot 5^4 \cdot 7^2 \cdot 11 \cdot 13 \cdot 23 \ = \ 8\ 315\ 553\ 613\ 086\ 720\ 000.$$

To understand a little better why geometry suggested that such a simple group could be found, we shall look a little deeper into a couple of the familiar examples in two and three dimensions. Even then, however, we will not see the complete picture.

First, consider the equilateral triangle. Its symmetry group consists of the three rotations 0°, 120°, and 240°, together with three reflections, each about an axis through a vertex and the side opposite. This group of order six is not simple, since it contains a subgroup of order three ($6/3 = 2$ implies that it is normal). This subgroup, being of prime order, is simple. (Here we use the theorem of Lagrange, which states that for finite groups, the order of a subgroup must divide the order of the group. Since H has prime order, it can have no nontrivial subgroups. Clearly, then, it can have no nontrivial normal subgroups!) This same argument applies to all the regular polygons with a prime number of sides. When the number of sides is not prime, there is a "subconfiguration" in the polygon that is fixed as a whole by a normal subgroup. We illustrate this with the square. Connect vertices 1 and 3 with a diagonal. The elements of the group that fix the diagonal (as a whole) are e, g, f^2, g, and f^2. These form a normal subgroup of the symmetry group. As it turns out, it is not simple either, and the presence of the symmetrical diagonal rules out large (relatively) simple groups from being associated with D_4.

Second, we move to three dimensions and consider the regular tetrahedron. As with the triangle, there is a lot of symmetry. Unfortunately, even with all that symmetry, the corresponding symmetry group has normal subgroups that in turn have normal subgroups. The problem in this case is that other geometrical considerations take over. Let us explain. Connect the midpoints of opposite edges with line segments. These three line segments not only are perpendicular to the edges they connect but are in fact mutually perpendicular and are connected together at their midpoints. This six-pointed "star" is illustrated in figure 9.9. The presence of this "star" configuration suggests that the symmetry group will have a nontrivial normal subgroup fixing that configuration (which it does). Of course, all this is in some sense very subjective. In general we ask, "Is there a lot of symmetry and no obvious subconfiguration?" As it turns out, the regular icosahedron, even though it does not possess as much symmetry as the tetrahedron, has no such subconfiguration, and its symmetry group contains a large simple subgroup.

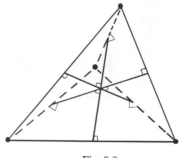

Fig. 9.9

Much the same happens with polytope P, the figure in twenty-four-dimensional space. It possesses tremendous symmetry. For example, each vertex is adjacent to 4600 other of its 196 560 vertices, each adjacent pair is adjacent to 891 other vertices, and so on. This, together with the absence of any subconfiguration, strongly suggested the presence of nontrivial normal subgroups (Thompson 1983, 109–18).

Of course, the geometry is no longer easy. However, without it, the conjectures would not have been made. The mathematical literature contains many more geometric connections with groups, and new connections are continually being made. We have only scratched the surface.

In summary, we have looked at three distinct uses of geometry today. First we considered affine geometries and their use in designing statistical experiments. Then we gave a novel example of the use of symmetries in archaeology. Finally, we gave a glimmer of how algebraists use geometric figures to help make conjectures about groups. Where will geometry be used next?

REFERENCES

Beck, Anatole, Michael N. Bleicher, and Donald W. Crowe. *Excursions into Mathematics.* New York: Worth, 1969.

Chapman, Kenneth M. *The Pottery of San Ildefonso Pueblo.* School of American Research Monograph No. 28. Albuquerque: University of New Mexico Press, 1970.

Crowe, Donald W. "The Geometry of African Art II: A Catalog of Benin Patterns." *Historia Mathematica* 2 (1975): 253–71.

Gardner, Martin. "How Three Modern Mathematicians Disproved a Celebrated Conjecture of Leonhard Euler." *Scientific American,* May 1959, pp. 181–88.

Thompson, Thomas M. *From Error-correcting Codes through Sphere Packings to Simple Groups.* Carus Mathematical Monographs 21. Washington, D.C.: Mathematical Association of America, 1983.

Washburn, Dorothy K., and R. G. Matson. "Use of Multidimensional Scaling to Display Sensitivity of Symmetry Analysis of Patterned Design to Spatial and Chronological Change: Examples from Anasazi Prehistory." In *Decoding Prehistoric Ceramics,* edited by Ben A. Nelson. Carbondale, Ill.: Southern Illinois University Press, 1985.

Zaslavsky, Claudia. *Africa Counts.* Boston: Prindle, Weber & Schmidt, 1973.

10

Geometry—a Square Deal for Elementary School

Marcia E. Dana

Too often geometry has been viewed by elementary school teachers as simply the study of such terms as *rectangle, line segment, angle,* and *congruence.* Kindergarten teachers have taught shape recognition (circles, squares, and triangles) much as they do letter and number recognition. Even in the intermediate grades, geometry has often been neglected until the end of the year when a few pictures and terms have been quickly introduced and drilled.

Teachers' decisions about what geometry to teach have been heavily influenced by what geometry they had (usually a smattering throughout middle school followed by a tenth-grade geometry course that focused on terms and proofs), what is included in current textbooks (very little), and what is tested on the achievement tests at their level (not much). The message from all these sources has usually been the same: geometry is boring, unimportant, irrelevant, and inappropriate for elementary school.

As a teacher, I have included more geometry each year because I am convinced that geometry is a great untapped source of ideas, processes, and attitudes that are entirely appropriate for elementary school. I have observed repeatedly that geometry can be exciting, motivating, rewarding, thought-provoking, and sometimes challenging (often as much so for the teacher as for the student). Students who are not whizzes at arithmetic are often the first to solve a puzzle, the most artistic in creating designs, and the most persistent when asked to find all possible patterns or shapes of a given kind. Children who are "good arithmetic students" often want to be told how many possibilities there are or to figure out the answer mentally instead of experimenting and using trial and error.

The geometric activities that follow are ones I have used with first-through fifth-grade children. (I have been fortunate enough to write elementary school curriculum, and this has given me time to think about many geometric activities.) They are quite different from those that appear in most elementary texts but are easy to use and provide enjoyable learning experiences for children.

Although squares are the focal shape of each of the puzzles, searches, and patterns that follow, many concepts are involved, including angles, symmetry, sides, congruence, diagonals, and midpoints. I have used squares to show the variety of ideas that can stem from just one shape, but the activities can be done with other shapes (triangles, rectangles, hexagons, circles, etc.). You can devise variations centering on other shapes and properties.

Puzzles

The basic idea of these puzzles is to make a certain shape or design with the given pieces. In all the suggested puzzles, either the shape to be made is a square or the pieces begin as a square. While the children are doing the puzzles, they are finding out about the relationships of the parts of a square to the whole square, of acute angles to right angles to obtuse angles (minus the vocabulary), and of the diagonals to the sides of the square. They are learning that a shape stays the same no matter how it is turned, flipped, or slid. They learn that it might be necessary to try a variety of ways before they find one that works. Often they develop a willingness to persevere because they are not expected to already know how to solve the puzzle.

Cutting Puzzles

A series of puzzles follows, ranging from simple to more complex. It is hard to attach a grade-level label because a puzzle that is easy for a particular first grader may be difficult for a particular fourth grader. The puzzles differ in the number of pieces (fewer pieces usually make an easier puzzle) and the relationships between the pieces (obvious relationships are usually easier). The pieces, along with an outline of the square, may be supplied to young children. The children can then move the pieces around on the outline until it is filled. Older children can just be given the pieces and asked to make a square.

One way to organize a square-puzzle activity is to make squares of various colors of construction paper or tagboard and cut each square a different way. Each puzzle can then be put into an envelope, and children may try as many as they wish. Another way is to make a ditto showing the separate pieces to a puzzle. The children then cut out the pieces, put them together to make a square, and then glue them in their square shape on to another piece of paper. Figure 10.1 shows some possible ways to cut squares for puzzles. *Note:* Puzzles are shown on grid paper for clarity. Make the puzzles on construction paper or tagboard, however, not on grid paper.

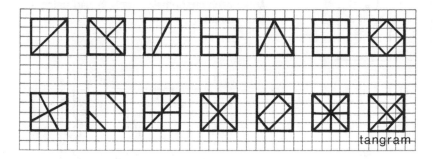

Fig. 10.1

There are many variations on the basic idea of cutting a square into pieces and then making other shapes from those pieces. Here are three examples. You can make them yourself or have older children make them for you (fig. 10.2–10.4).

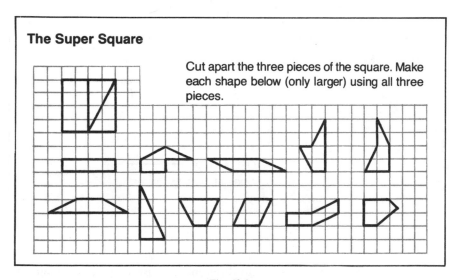

Fig. 10.2

Note: Use a square large enough for children to handle. Use grid paper to trace the shapes to be made on plain paper.

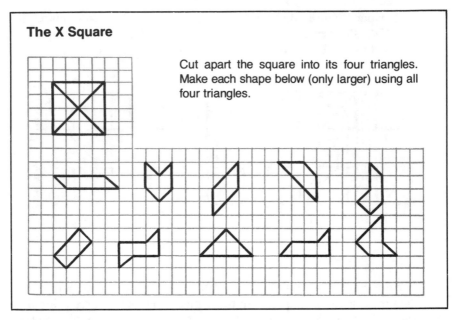

The X Square

Cut apart the square into its four triangles. Make each shape below (only larger) using all four triangles.

Fig. 10.3

Note: Use a square large enough for children to handle. Use grid paper to trace the shapes to be made on plain paper.

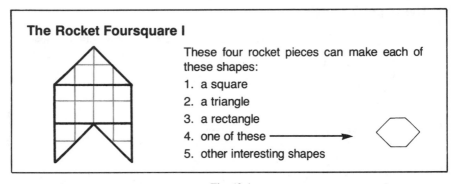

The Rocket Foursquare I

These four rocket pieces can make each of these shapes:
1. a square
2. a triangle
3. a rectangle
4. one of these ———————→
5. other interesting shapes

Fig. 10.4

Folding Puzzles

The next type of puzzle is quite different because its pieces remain attached and it is folded to make new shapes. Folding puzzles can range from simple to quite complex. One of their advantages is that there are no small pieces to cut or lose. A possible disadvantage is that a child must be able to fold carefully and only on the lines (in either direction). To make these puzzles, begin with a square that measures approximately 5 inches (fig. 10.5).

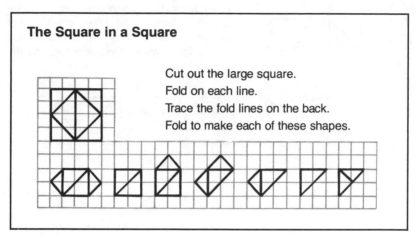

Fig. 10.5

This puzzle can be used with first graders who can handle the folding and also serve as an introductory folding puzzle for many older children. I usually give it in the form of a ditto with written directions and small pictures of the shapes to be made. Other somewhat more complex folding puzzles follow (figs. 10.6 and 10.7).

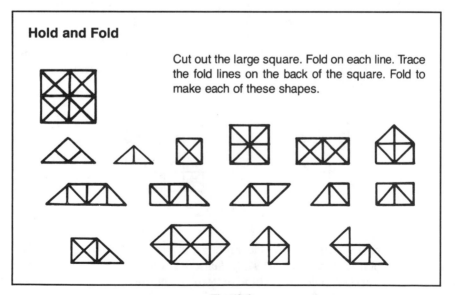

Fig. 10.6

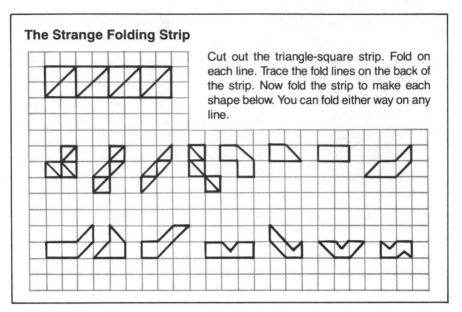

Fig. 10.7

Note: Trace these shapes on plain paper. Use a larger grid paper.

The last two shapes in the hold and fold puzzle are hard problems, but I promise they can be done. Adults in workshops often insist that they are impossible, but children usually experiment until they get them (or until they finally give up). Notice that the inside lines do not appear on most of the shapes to be made. This makes the shapes more challenging. I usually have children draw in the lines to show how they arrived at their solutions.

Lots of other folding puzzles can easily be devised. A few other suggestions appear in figure 10.8.

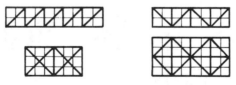

Fig. 10.8

An interesting variation on the folding puzzle idea involves the use of color and color patterns (fig. 10.9). (*Alternative:* I usually have the children make up several of their own color patterns using the accordion strip and then suggest they challenge a friend to solve them.)

Accordion Squares

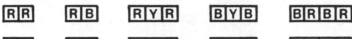

Color the strip of squares (B = blue, G = green, R = red, and Y = yellow). Then cut out around the strip. Fold on the lines between the squares. Color the back of each square the same color as the front. Fold your accordion to make each of these combinations.

Fig. 10.9

Accordion Triangles

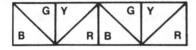

This puzzle can make all the shapes and patterns below and many more.

Fig. 10.10

Pinwheel Square

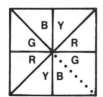

Cut on the dotted line. Fold to the shapes below.

Fig. 10.11

Folding color puzzles can be simple or more difficult, depending on how many colors and parts the puzzle has. Shape and color can be combined, as in the two puzzles shown in figures 10.10 and 10.11.

As I mentioned before, squares are only one shape from which any of these puzzles can be made. Other especially productive shapes include equilateral triangles, rectangles, hexagons, and circles.

Searches

"How many different ways can you color this design?" or "How many designs can you make using these dots?" are typical ways a search problem begins. Such problems require students to organize their separate solutions and to look for patterns in their solutions that will help them discover all the possibilities. They must also be able to recognize and discard duplicates.

One major point you must clarify for students before they begin a search activity is what the term *different* means for the particular activity. For example, in an activity showing sixteen of these squares, the directions say—

 Use one crayon and color these so that all 16 are different. In each square you can color in any arrangement of small triangles.

In this activity, *different* can include the orientation of the designs. In other words, designs that are congruent but turned in different directions are considered different for this activity. Thus, these are all different:

The sixteen solutions, then, are shown in figure 10.12:

Fig. 10.12

If orientation were not considered "different," there would be only six solutions:

My experience has been that students, even at first- or second-grade level, understand both meanings of "different" as long as you give examples. Usually I use the number of solutions to determine whether to include orientation as being different. In the next coloring search, I choose to say that orientation doesn't make a different design, since there would be too many solutions for most children to find if all were included. The activity can include twenty-four of these rectangles (or fewer if you prefer).

The rectangle produces twenty-four different designs if orientation doesn't count—sixty-four different designs if it does. Some students will need proof that two designs, such as those in figure 10.13, are not different. They can cut out both designs, then flip one over and put it on top of the other so the designs coincide. Or I might help them describe the designs verbally: "One in a corner of one row, and two in the opposite corner of the other row." Some children will then see that both designs fit that same description.

Fig. 10.13

Here are some other coloring searches to try (fig. 10.14). Again, shapes other than squares can be used.

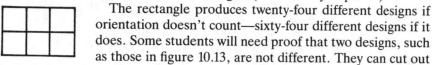

Fig. 10.14

Variations on search activities can include drawing instead of coloring, as in figures 10.15–10.17.

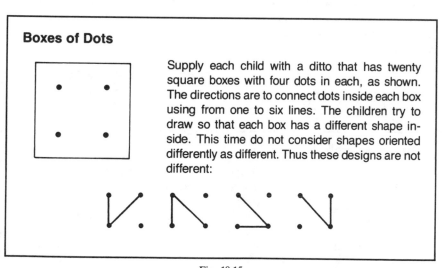

Boxes of Dots

Supply each child with a ditto that has twenty square boxes with four dots in each, as shown. The directions are to connect dots inside each box using from one to six lines. The children try to draw so that each box has a different shape inside. This time do not consider shapes oriented differently as different. Thus these designs are not different:

Fig. 10.15

There are 18 possible answers for this search.

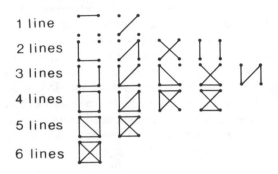

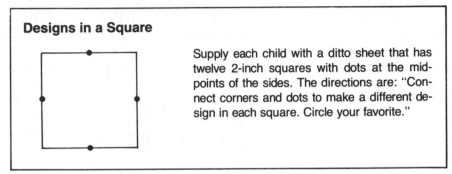

Designs in a Square

Supply each child with a ditto sheet that has twelve 2-inch squares with dots at the midpoints of the sides. The directions are: "Connect corners and dots to make a different design in each square. Circle your favorite."

Fig. 10.16

Orientation should probably not be considered different for the search in figure 10.16. I ask children to copy their favorite design onto a larger square and then color it. Many interesting and colorful designs result. Here are some samples:

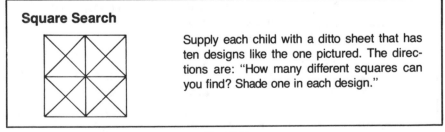

Square Search

Supply each child with a ditto sheet that has ten designs like the one pictured. The directions are: "How many different squares can you find? Shade one in each design."

Fig. 10.17

There are ten different squares to be found in figure 10.17:

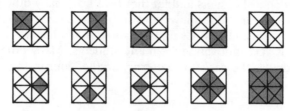

Patterns

Continuing a pattern or devising an original one can involve many processes and ideas: organizing, describing, logic, congruence, order, symmetry, spatial relationships, and creativity. Patterns that focus on squares also illustrate the properties of squares. Students can copy patterns, continue patterns, or create their own patterns. The patterns can range from quite simple to extremely complex. Patterns need not be only linear. Any design that repeats in some way can be thought of as a pattern. Wallpaper repeats itself in more than one direction and often reveals a complex pattern.

Some simple patterns involving squares can be continued by the student (fig. 10.18).

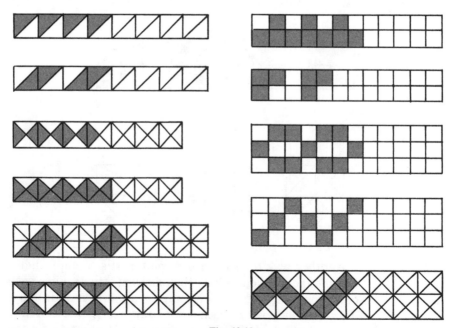

Fig. 10.18

Also, students can make their own patterns from unshaded designs. Provide several identical designs and let them see how many different patterns they can make. Figure 10.19 shows several from the same design, and there are many more possibilities. Using two colors is another way to create new patterns and is very motivating for many students. Many different designs can be provided as the basis for patterns, and, of course, they need not be made of squares.

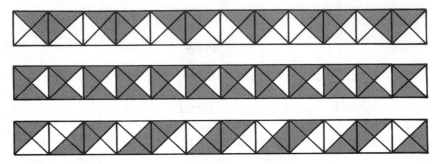

Fig. 10.19

Patterns need not be linear. Students can color a design of four squares in many different ways using two colors to create patterns.

Other designs can be used as well:

 Another way to learn about patterns is to copy them. Fourth-grade students enjoy the following activity. Give each one a sheet of paper with a large square drawn on it and a dot at the midpoint of each side. Draw a large one carefully on the chalkboard and give directions a step at a time for the design in figure 10.20. They follow using a ruler as carefully as possible. Here are the directions:

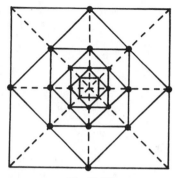

1. Draw light lines connecting opposite corners and opposite dots. These will be erased when the design is finished.
2. Connect each dot to the one next to it to make a square.
3. Draw dots where the lines you drew cross the light lines. Then connect those dots to make a square.
4. Continue. Draw as many squares as you can. Erase the light lines.
5. Name the design and color it.

Fig. 10.20

The difference between individual children's ability to see the pattern and follow it is quite pronounced. Some continue on their own after seeing the first two squares made. Others need to be led line by line through the whole pattern. Once the pattern is copied, the children can color them. Every child's pattern is usually colored differently (and in ways the teacher has never considered). Sometimes children want to copy this pattern several times, and their fascination and enthusiasm can continue for weeks.

A Square Deal

Along with the direct benefits of the ideas, processes, and concepts involved in teaching geometry in elementary school come several hidden benefits. Geometry provides a great change of pace from arithmetic and involves different types of thinking. Since the best puzzle solvers and "searchers" often are children who don't shine in arithmetic, these activities give such children a chance to be praised and rewarded for their accomplishments. Other children will gain more respect for them. Perhaps one of the best hidden benefits of teaching geometry is that it's fun for the children and can be for you as well. You are cast more in the role of observer and even fellow learner during many of the activities.

Geometry need not be taught as a whole unit once a year. Instead, try using one activity every day or at least twice a week throughout the year. The children will be getting more of a "square deal" mathematically, and you may be inspired to create more of your own geometry activities.

Spatial Perception and Primary Geometry

John J. Del Grande

$\mathbf{P}$RIMARY children enter kindergarten with many intuitive notions of space. A large portion of infant and early behavior is essentially "spatial" in character because it is prelinguistic, since a child's first encounters with, and explorations of, the world are undertaken without the benefit of language. In this period children's thinking is dominated by the interpretations they give to their experiences of seeing, hearing, touching, moving, etc., that is, their perceptions of space.

Spatial perception is the ability to recognize and discriminate stimuli in and from space and to interpret those stimuli by associating them with previous experiences. Eighty-five percent of all information coming into the body from the environment enters through the visual system, and vision is developed as a result of many cumulative experiences. The study of perception is rooted in psychology, philosophy, and physics; this may, therefore, account for the fact that a universally agreed-on definition does not exist and can change according to the task at hand.

The lack of a universal definition should not keep us from considering the relationships of spatial perception and geometry. As Hoffer (1977, p. 92) summarized:

> It appears that visual perception skills and geometry concepts can be learned simultaneously, since geometry requires that the students recognize figures, their relationships, and their properties. Informal geometry could easily be taught and included with a visual perception training program so as to improve students' visual perception.

The very nature of the mathematical activities involved in primary geometry makes them ideal vehicles for acquiring spatial perception experiences and gives teachers an excellent opportunity to observe and detect children's perceptual problems at an early age. Thus, a clear understanding of spatial perception abilities should make it possible to design geometry programs and select mathematical activities that will improve students' visual perception (Del Grande 1986).

The suggestions in this article differ from the usual direction of investigations where studies are concerned with what spatial perception abilities are necessary for success in geometry. Here we shall consider what geometric activities might enhance and develop a primary child's spatial abilities.

Spatial Perception Abilities

Psychologists were perhaps the first to concern themselves with the study of the interaction of the perceiver with his or her environment. Piaget's investigations and theories have contributed extensively to what is known about a child's conception of space geometry and geometric transformations (Piaget and Inhelder 1967). But Piaget was not, on the whole, concerned with defining constructs and enumerating abilities. Frostig and Horne (1964) were early pioneers in the identification and testing of perceptual abilities. In her book *The Visual World of the Child,* Eliane Vurpillot (1976) not only provides a comprehensive account of the growing field of research in spatial perception but also attempts to integrate the different points of view that have been recorded.

Frostig and Horne (1964) produced test materials for the first five of the seven spatial abilities listed below, and Hoffer (1977) examined two more abilities, namely, visual discrimination and visual memory. These abilities seem to have greatest relevance to academic development:

1. Eye-motor coordination
2. Figure-ground perception
3. Perceptual constancy
4. Position-in-space perception
5. Perception of spatial relationships
6. Visual discrimination
7. Visual memory

Eye-motor coordination is the ability to coordinate vision with movement of the body. Children who have difficulty with simple motor skills and movements have difficulty thinking of anything else as they concentrate on the task at hand. For example, if a child is having trouble joining dots on geopaper, arranging wooden blocks to make a solid, or using a ruler to draw a line, the effort alone may be enough to absorb the child completely. Only when such coordination is habitual will children be able to give their whole attention to the act of learning or to the perception of external objects, since movements need no longer require great mental concentration. This fact suggests that thinking and doing are separate acts.

Figure-ground perception is the visual act of identifying a specific figure (the focus) in a picture (the ground). In focusing attention on a figure, one must disregard the extraneous markings surrounding it and not be distracted by irrelevant visual stimuli. Vurpillot (1976) discusses the extent to which children four to seven years of age can break down perceptual units into their components and reassemble them in new forms as embedded figure problems. It has been found that 70 percent of four-year-olds can identify overlapping figures. Those children who do not have figure-ground perception to start are very likely to acquire it through appropriate intervention.

Perceptual constancy, or the constancy of shape and size, is the ability to recognize that an object has invariant properties such as size and shape in spite of the variability of its impression as seen from a different viewpoint. For example, a person with perceptual constancy will recognize a cube seen from an oblique angle as a cube, even though the eye gets a different image when the cube is viewed from squarely in front or directly above. Thus perceptual constancy assists a person in adjusting to the environment and stabilizes his or her world by emphasizing the permanent quality of objects rather than their continually changing appearance as they are moved relative to the observer. Frostig and Horne (1964) found that the development of perceptual constancy depends in part on learning and experiences that are provided by activities of a geometric nature.

Position-in-space perception is the ability to determine the relationship of one object to another object and to the observer. The lack of position-in-space ability results in reversals, which poses a dilemma for mathematics educators. On the one hand, we want children to see that two figures are the same (i.e., congruent) if one is the image of the other under a slide, flip, or turn. But on the other hand, we say that **b, d, p,** and **q** are all different. A focus on the motions that move one figure onto another may help children resolve this difficulty.

Perception of spatial relationships is the ability to see two or more objects in relation to oneself or in relation to each other and is closely related to position-in-space perception for some tasks. For example, a person with such an ability sees that figures are congruent to each other when one is the image of the other after a slide, flip, or turn.

Visual discrimination is the ability to distinguish similarities and differences between objects (Hoffer 1977). The activities of sorting and classifying objects and geometric shapes such as attribute blocks assist children in learning visual discrimination. The children can use pictures and abstractions as they develop their visual discrimination by making visual and verbal comparisons of the things they see.

Vurpillot cautions that the difficulties young children have in differentiating between orientations in activities involving several figures may be inherent in the task itself. Comparing several figures involves the exploration of all alternatives presented and then comparing them one by one with the original; this involves a strategy that a young child may not possess but can be taught.

Visual memory is the ability to recall accurately an object no longer in view and then to relate its characteristics to other objects either in view or not in view (Hoffer 1977). Most people retain small amounts of visual information—about five to seven items—for short periods of time. Hoffer states that to remember great amounts of information we must store the information in long-term memory through abstractions and symbolic thinking.

Growing children acquire spatial perception through experiences encountered in their environment. Spatial perception not only helps children get to school but is essential in enabling them to read, write, spell, do arithmetic and geometry, paint, play sports, draw maps, and read music.

Spatial Perception Activities

In a balanced primary program, mathematics should be integrated with other activities of the child's day at school. It is not surprising to find geometry related to art, geography, and even reading and writing. The activities in this section can be modified for the level of the student involved, and suggestions will be made throughout to help achieve this end. Verbal instructions are frequently necessary for young children.

Eye-hand coordination can involve activities such as drawing with and without guidelines and tracing and coloring as illustrated in the two examples in figure 11.1.

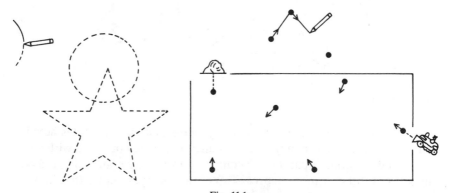

Fig. 11.1

Teachers can design many similar activities where the effort required is suitable for children of a given grade level. For example, joining two dots may be a good exercise for children in grade 1, whereas joining many dots would be more appropriate for children in grade 3. Instructions should be verbal when necessary, and a discussion with student-created "stories" should be encouraged.

Figure-ground perception is an important ability that, if not acquired, can handicap children. Activities in figure-ground perception include those that involve intersecting lines, intersecting figures, hidden figures, overlapping figures, figure completion, figure assembly, similarities and differences, and the reversal of figure and ground. Two sample activities are shown in figure 11.2.

- Draw a blue path around the rectangle.   Finish the figure in B to look like A.
- Color inside the triangle.

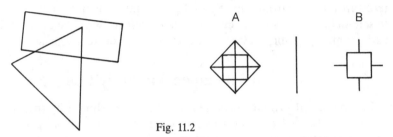

Fig. 11.2

Figure-assembly activities involve filling a figure using a number of geometric figures such as those found in parquetry blocks, pattern blocks, tangrams, or felt cutouts. For example, the students are given five figures made from felt or cardboard, as shown on the left in figure 11.3, and are required to fill the outline of the figure on the right. This activity can be made more difficult by giving children more pieces than necessary to fill the figure and by giving them the instruction, "Use as many pieces as you need to fill the figure."

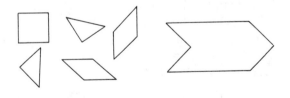

Fig. 11.3

Perceptual constancy, or constancy of shape and size, includes activities that involve shape constancy, size constancy, constancy of shape with figure-ground perception, apparent size compared to real size, and comparison of the sizes of three or more figures. Two of the many examples that illustrate perceptual constancy are given in figure 11.4.

- Trace all the squares in blue

- 3 figures below can be fitted together to make a square like the one in the box
- Mark the pieces with an X

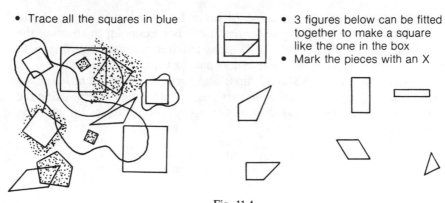

Fig. 11.4

Position in space activities include reversals and rotations of whole figures, change of position of detail, and mirror patterns. Two examples are given in figure 11.5.

- Look at the figure in the box
- The picture shows the figure turned in different ways
- Find where the black triangle belongs Color it

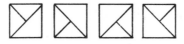

- Look at the letters in the box
- Color them
- Color all letters below that are the same as the red one
- Color all the rest the same way

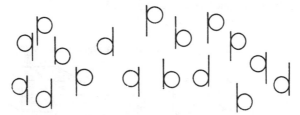

Fig. 11.5

While children are doing some of these activities, it is interesting to observe the problem-solving strategies used. For example, in rotating the square in the first activity (fig. 11.5), some children may use their fingers to place the black triangle, whereas others may just turn the page.

Activities that involve slides, flips, and turns, such as identifying the motion, identifying images, drawing images, and drawing patterns using the motion, could also be included in this category. Two examples are given in figure 11.6.

- Look at the picture of the comb in the box
- Find combs that are *slide images* of the comb in the box
- Circle them

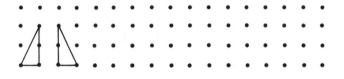

- The first two figures of a pattern are shown
- Use flips to draw three more figures in the pattern

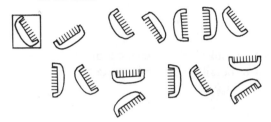

Fig. 11.6

Perception of spatial relationship activities include relating the position of two or more objects, noting similarities and differences, finding the shortest path to a goal, completing a figure, connecting dots, completing a sequence, and assembling parts. An early activity might involve replicating a cube structure such as the one shown in figure 11.7.

Place 3 cubes on a table as shown below.

Fig. 11.7

- "Put your cubes together like mine and then place them on top of my cubes" (no color match is required).
- "Put your cubes together like mine and then place them beside my cubes."
- Give the child more cubes so that he or she can now match colors and then repeat the previous activities.

Similar activities can be designed using colored rods, Popsicle sticks, beads, or pattern blocks. Two other activities for the perception of spatial relationships appear in figure 11.8.

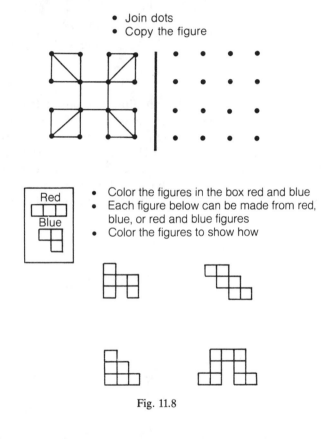

Fig. 11.8

Visual discrimination involves identifying a pair of objects that are the same, a pair of objects that are different, one object that is different from the others, and several objects that are the same but different from the others. Two activities exemplifying this spatial ability are shown in figure 11.9. The activities with figures that are "the same" also involve constancy of shape and visual memory.

- One figure is different from the rest
- Circle the figure

Two figures are exactly the same. Circle them.

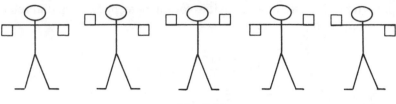

Fig. 11.9

Visual memory activities involve recalling one object from two or more objects, one object with reversals, or the position of many objects. Further activities also include drawing figures from memory or completing a figure from memory. For example, in figure 11.10, children are shown a picture of a shelf with toys and then asked to remember the placement of the toys after the picture is hidden. This activity can be made easier by reducing the number of toys used or harder by adding toys that do not belong.

- Draw a line from each toy to the shelf on which it belongs

Fig. 11.10

Another set of useful activities consists of (1) showing a figure drawn on dot paper for ten seconds, (2) hiding the figure, and (3) asking children to reproduce the figure from memory on dot paper. Such activities can be designed to be very easy or very hard by varying the complexity of the figure to be remembered and changing the number of dots used.

Implications

Perceptual abilities are important for beginning school success and have a strong influence on the stability of the child. Early school adjustment and progress are enhanced if a child's spatial perception abilities are equal to the tasks the child encounters. The visual world is the result of a slow process that creates a world of objects, similar or different, interdependent, and meaningful.

Builders of geometry curricula should take into account the development of the child's understanding of space and processing of visual information. Those involved with the teaching of primary mathematics must be aware of the spatial abilities of the students they teach and attempt to adjust instruction to those abilities. As research provides more information, those teachers will be able to do a more effective job of adapting instruction to the needs and abilities of their students.

Spatial perception activities will be effective only if they are integrated into a well-rounded program and take into account the child's total development.

REFERENCES

Del Grande, John J. "Can Grade Two Children's Spatial Perception Be Improved by Inserting a Transformation Geometry Component into the Mathematics Program?" Doctoral dissertation, Ontario Institute for Studies in Education, 1986.

Frostig, Marianne, and David Horne. *The Frostig Program for the Development of Visual Perception.* Chicago: Follett Publishing Co., 1964.

Hoffer, Alan R. *Mathematics Resource Project: Geometry and Visualization.* Palo Alto, Calif.: Creative Publications, 1977.

Piaget, Jean, and Bärbel Inhelder. *The Child's Conception of Space.* New York: W. W. Norton & Co., 1967.

Vurpillot, Eliane. *The Visual World of the Child.* London: George Allen & Unwin, 1976.

Similarity: Investigations at the Middle Grades Level

Alex Friedlander
Glenda Lappan

THE famous Dutch mathematician Hans Freudenthal describes geometry as experience with, and interpretation of, "the space in which the child lives, breathes, and moves" (1973, p. 403). From this perspective we can think of children beginning to learn geometry as soon as they are able to see, feel, and move in the space they occupy. As children grow, they begin to perceive characteristics of the objects in that space, such as shape, size, position, motion, order, and growth. Our task as teachers of geometry is to provide the kinds of experiences for students that will enhance their understanding of the space about them.

Teaching the concept of geometrical similarity will be used in this article as an example of teaching geometry through informal, exploratory activities that employ concrete materials. Most of the ideas described here are taken from an instructional unit developed by the NSF-funded Middle Grades Mathematics Project (Lappan et al. 1986).

Why Teach Similarity?

We chose this topic to illustrate some principles in teaching geometry at the middle grades level, since the acquisition of the concept of similarity is important to the development of children's geometrical understanding of their environment and of proportional reasoning. Phenomena that require familiarity with enlargement, scale factor, projection, area growth, indirect measurement, and other similarity-related concepts are frequently encountered by children in their immediate environment and in their studies of natural and social sciences. Fuson points out the need for instruction in similarity (1978, p. 259):

> Similarity ideas are included in many parts of the school curriculum. Some models for rational number concepts are based on similarity; thus, part of students' difficulty with rationals may stem from problems with similarity ideas.

Ratio and proportion are part of the school curriculum from at least the seventh grade on, and they present many difficulties to the student. Standardized tests include many proportion word problems. Verbal analogies (a:b::c:d) form major parts of many intelligence tests. Similar geometric shapes would seem to provide a helpful mental image for other types of proportion analogy situations.

Identifying Similar Shapes

Two informal tests for similarity can be explored to build basic notions of similarity: (1) the overhead projection test and (2) the diagonal test.

Overhead Projection Test

Children will quickly recognize that the shadow of a shape placed on the overhead projector is similar to the original figure.

Therefore, whenever we have to consider whether two cut-out shapes are similar (i.e., one is the enlarged version of the other), we place the small figure on the projector and attach the larger one to the screen; by moving the projector back and forth, we attempt to match the larger figure with the projected shadow. Even though the projector can always be moved so that at least one edge of the figure image matches an edge of the figure attached to the screen, it is only when both entire figures can be made to match that they are similar. Several activities make use of this overhead projection strategy for determining similarity.

1. Have the children apply the overhead projection test to different pairs of rectangles.

2. Have them create a sequence of rectangles by folding a sheet of paper in half crosswise and cutting it on the fold line; repeat this step six to seven times, each time with one of the newly created rectangles (fig. 12.1). Then have them categorize the resulting rectangles into groups of similar rectangles using the overhead projector.

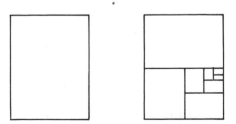

Fig. 12.1. Creating a sequence of rectangles

3. Use the projection test for other than rectangular shapes.

4. Use the projection test for different pairs of circles. Ask the students what they can conclude.

Diagonal Test

Children can be led to discover that the diagonals of similar rectangles fall in a straight line when the rectangles are nested. This gives a quick visual way of determining whether two rectangles are similar.

Have the children apply the diagonal test to the collection of rectangles from part 2 of the overhead projector activities:

The Diagonal Test

Take a rectangle from your collection. Place it so that the longer edge is at the bottom, and draw a diagonal line from the lower left corner to the upper right. *(If children have difficulty with directions, simply have them draw both diagonals.)* Nest your collection of rectangles in order of size, the longer edge always at the bottom and with the lower left corners aligned (fig. 12.2).

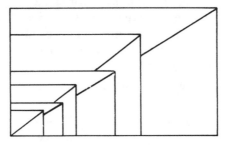

Fig. 12.2. The diagonal test

In part 2 of the overhead projector activities, the children will already have decided that the rectangles belong to two families of similar rectangles. Here they will see that the diagonals of each family align.

Creating Similar Shapes

Intuitive notions about similarity can be strengthened and expanded by introducing children to some techniques that enable them actually to *create* similar shapes. Two of these techniques are based on the principle of point projection: (1) the variable tension proportional divider (VTPD) and (2) the constant tension proportional divider (CTPD).

Variable Tension Proportional Divider

This is an intriguing way to enlarge a figure that gives reasonable results in the hands of most children. The final figures are not perfectly accurate, but they do clearly convey the intuitive "same shape" notion of similarity. The VTPD allows children to make enlargements of irregular figures, figures with curved edges, or simple pictures from magazines or cartoons.

1. Have them build a VTPD by knotting together two identical rubber bands (#18's work well). Use masking tape to attach a large sheet of paper to the desk. Choose an anchor point to the left of the figure to be enlarged so that the knot will be taut on all parts of the figure to be enlarged.

2. Have them put one end of the rubber band at the anchor point and the pencil in the other end. Holding the anchor end firmly, they move the pencil as the knot traces the figure. They must keep their eyes on the knot while the pencil hand does the drawing. See figure 12.3.

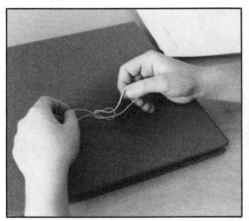

Fig. 12.3. Preparing and working with a VTPD

3. Ask what happens if you change the anchor point to another place.
4. Ask what happens if you knot three rubber bands in a row and let the first knot trace the figure.

Constant Tension Proportional Divider

Using paper strips (which have a "constant tension"), children can very precisely enlarge simple geometric figures composed of straight edges. (The technique is cumbersome for figures that have curved edges.) We shall illustrate making an enlargement that has linear dimensions double those of an original triangle *ABC*. (For young children, you may need to present this activity in somewhat simpler terms than those used here.)

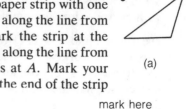

(a)

1. Cut out several stiff paper strips about a half-inch wide and eleven inches long. Mark an anchor point *P* for the enlargement (fig. 12.4a). Position the paper strip with one of its ends at vertex *A* and with its edge along the line from the anchor point to the vertex *A*. Mark the strip at the anchor point (fig. 12.4b). Slide the strip along the line from *P* to *A* until the mark that was at *P* is at *A*. Mark your paper at the point *A'*, which is now at the end of the strip (fig. 12.4c). This is the position of the vertex of the new triangle. Notice that *A'* is twice as far from *P* as *A* is from *P*.

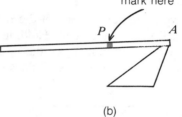

(b)

2. Repeat this process with *B* (i.e., find a point *B'* twice as far from *P* as *B* is from *P*) and then with *C* (fig. 12.4d).

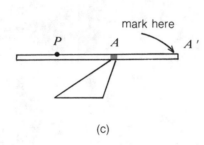

(c)

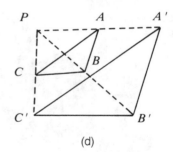

(d)

Fig. 12.4. Projecting with a CTPD

3. What happens if you move the anchor point to another place?
4. What happens if you mark *A'*, *B'*, and *C'* three times as far from the anchor point?
5. What happens if you mark *A'*, *B'*, and *C'* only one-half as far away from *P* as *A*, *B*, and *C* are? (Fold the paper strip to find half distances.)

Area Growth

The principle of area growth presents many cognitive difficulties to children of this age and even to high school students and adults. The area growth of similar shapes requires the recognition of the (somewhat coun-terintuitive) fact that the enlargement of a figure by a linear scale factor of *n* will increase its area by a factor of n^2. For example, if a rectangle 3 cm by 4 cm is enlarged into a rectangle 9 cm by 12 cm, the linear scale factor is 3 but the new area is 9 times as large as the original area.

Two activities have been designed to present the area relationship in a visual manner: "rep-tiles" and everyday applications.

Rep-tiles

Exploring figures that replicate in specific ways gives an informal setting for beginning notions about how area grows. The following activity is based on an idea from Martin Gardner's book *The Unexpected Hanging* (1969; chap. 19, "Rep-tiles: Replicating Figures on the Plane").

1. Cut out four exact copies of each of these shapes (fig. 12.5).

Fig. 12.5. Rep-tiles

Try to put the four figures of a kind together to form a figure that is *similar* to the original. This is called "rep-tiling" (fig. 12.6).

Fig. 12.6. Rep-tiling: Constructing similar figures

2. Draw a triangle and cut it out. Cut out three more copies of your triangle. Can you rep-tile with your four triangles? Can a rep-tile be formed from any triangle?

3. What about rectangles? Do they always rep-tile?

4. In a rep-tile made of four small figures, how does an edge of the rep-tile compare to an edge of the small figure?

5. Cut out nine copies of a triangle. Put all nine together to form a rep-tile. Remember that the big triangle must be similar to the original small triangles. Measure the edges of the small and large triangle. How do they compare?

Through this exploration, children see examples of an area growth factor of 4 (or 9) in a figure that has dimensions twice (or three times) the original.

Everyday Applications

The principle of area growth is frequently encountered in everyday situations, such as predicting the cost of painting a given area or the amount of material needed to build a reduced or enlarged version of a given shape. The following two problems are examples of such situations:

1. It costs $30 for paint to cover a blackboard at school. How much would it cost for paint to cover a board half that long and half that wide?

2. A package of rectangular paper sheets weighs 3 lb. What will be the weight of a package (same number of sheets and same quality of paper) with sheets twice as long and twice as wide?

The quantities involved in these situations depend on area. However, the questions require an understanding of area growth relationships and not just area measurements.

We suggest that initially the problems be posed in the same direct and abstract way as presented here. Usually the children will make the mistake of using the side scale factor in their answers (e.g., $15 for paint for the smaller blackboard).

Next, the children are asked to present the corresponding reduction or enlargement in a concrete way by actually cutting out paper rectangles. For example, by cutting out rectangles they should be able to show the smaller board and see that it is one-fourth, not one-half, of the larger board in area.

At this point, students should become aware of a discrepancy between the answer they have given and their own concrete representation of the problem. If this does not occur, the teacher must ask more leading questions to make it happen.

Conclusion

The activities described here provide the child a first encounter with the concept of similarity and are presented at the first two van Hiele levels of geometrical understanding. In a spiral curriculum, a more systematic approach may be employed the second time around: measurements, ratios, and proportions should emphasize the numerical aspects and the general principles involved in similarity.

We believe that only with such a background does the average student have any chance to understand the Euclidean approach to similarity at the high school level or to cope with real-life situations that involve the concept of geometrical similarity.

REFERENCES

Freudenthal, Hans. *Mathematics as an Educational Task.* Dordrecht, Netherlands: D. Reidel Publishing Co., 1973.

Fuson, Karen C. "Analysis of Research Needs in Projective, Affine, and Similarity Geometries, Including an Evaluation of Piaget's Results in This Area." In *Recent Research Concerning the Development of Spatial and Geometric Concepts,* edited by Richard Lesh, pp. 243–60. Columbus, Ohio: ERIC/SMEAC, 1978.

Gardner, Martin. *The Unexpected Hanging and Other Mathematical Diversions.* New York: Simon & Schuster, 1969.

Lappan, Glenda, William Fitzgerald, Elizabeth Phillips, and Mary Jean Winter. *Similarity and Equivalent Fractions.* The Middle Grades Mathematics Project Series. Menlo Park, Calif.: Addison-Wesley Publishing Co., 1986.

13

Visualizing Three Dimensions by Constructing Polyhedra

Victoria Pohl

THE best way to learn to visualize three dimensions is to make objects that demonstrate the spatial concepts. Students can observe and use many spatial relationships while they construct polyhedra. Attractive visual aids also stimulate creative thinking.

There seems to be little connection between the age of students and their ability to comprehend spatial relationships. Hence, though the classroom activities described here were created primarily for junior high school students, they are also suitable for grades 7 to college. Working with polyhedra, for instance, helps prepare high school students for the three-dimensional graphs encountered in college calculus. See Pohl (1986) for additional activities.

The activities also offer abundant opportunity for students to learn mathematical vocabulary and relationships. It is a good idea to introduce any unfamiliar words before the student begins work on an activity. The following words and phrases appear:

Meter	Octahedron
Centimeter	Cube
Midpoint of a line segment	Pyramid
Square	Vertex of a polyhedron
Equilateral triangle	Edge of a polyhedron
Vertex of a triangle	Parallel edges of a polyhedron
Center of a triangle	Skew edges of a polyhedron
Median of a triangle	Face of a polyhedron
Intersection of medians	Adjacent faces of a polyhedron
Side of a triangle opposite a	Interior of a polyhedron
vertex	Exterior of a polyhedron
Tetrahedron	

Following the activities are exercises that further strengthen students' spatial perception. The students should handle their own models while they discover and visualize various properties of three-dimensional space. These

144

exercises can be used for class discussion, as an assignment, or as enrichment material for a small group or an individual.

Activities

The materials needed are minimal and inexpensive. Each activity requires thin thread, a large needle, a metric ruler, scissors, and plastic straws. The needle must be large and the thread thin so the weight of the needle will pull the string through the straw. Colorful round plastic straws work best. The use of contrasting bright colors makes an attractive figure.

Activity 1. Tetrahedron   **Tetrahedron**

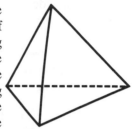

1. *Build the tetrahedron.* Choose six straws, all the same length. Thread the needle with about a meter of string. (Longer pieces tend to tangle; when the string becomes too short, tie on another piece.) Drop the threaded needle through three straws. Next tie the string to form a triangle, allowing no slack in the string (fig. 13.1). Drop the threaded needle through the fourth and fifth straws (fig. 13.2). Then drop the needle back through straw 2 and then through the last of the six straws (fig. 13.3). Pull the string tight. Finally drop the needle through straw 4 to return to the starting point (fig. 13.4). Tie the string securely. If you plan to firm up the vertices, proceed to step 2. Otherwise, cut off the ends of the string and consider the tetrahedron complete.

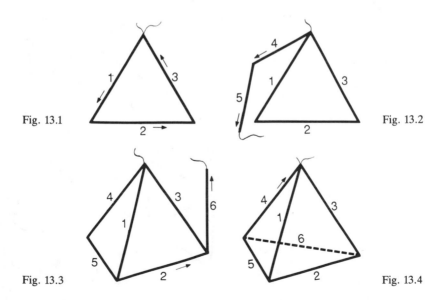

Fig. 13.1

Fig. 13.2

Fig. 13.3

Fig. 13.4

2. *(Optional) Stabilize the vertices.* Observe that three vertices have one pair of unattached straws like straws a and b shown here. To join the unconnected straws, tie a meter of string to one of the short strings on the tetrahedron, then thread it into a needle, and drop it through the straws until you have connected each straw to both of its neighbors. For example, passing the needle through straw a and then through b as shown will connect straws a and b. Pull and tie the string and cut off any excess.

Octahedron

Activity 2. Octahedron

Construct an octahedron by applying the techniques developed in Activity 1. Follow the steps below.

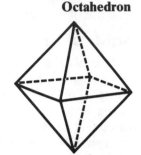

1. With twelve straws, all the same length, make four triangles and from them build two pairs of triangles (Fig. 13.5 and 13.6).

2. Connect and tighten straws 1, 2, and 3 (fig. 13.7), then straws 4, 5, and 6 (fig. 13.8).

3. (Optional) Stabilize the vertices.

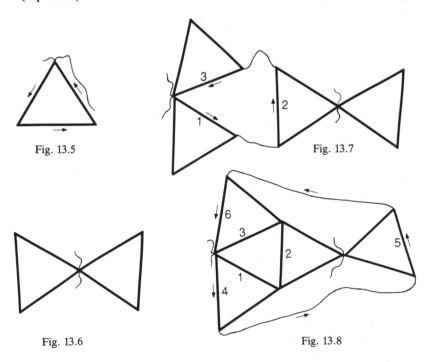

Fig. 13.5

Fig. 13.7

Fig. 13.6

Fig. 13.8

Activity 5. Tetrahedron inside Tetrahedron

**Tetrahedron inside
Tetrahedron**

1. Prepare the straws and build the large tetrahedron. Cut six drinking straws, each 13 cm long, and make a pinpoint hole at the midpoint of each. Then cut six cocktail straws, each 3.8 cm long. Use the 13-cm straws to build the tetrahedron. Follow both steps of Activity 1.

2. Stretch the medians. A rather thin needle is recommended for this step. Tie about a meter of string to any vertex of the tetrahedron. Thread the other end of the string into the needle. Figure 13.9 shows how to begin weaving the medians. First pull the string from the vertex (1) through the midpoint of the opposite side (2), then over to an opposite vertex (3), securing it by looping the threaded needle around the vertex and passing it through the midpoint of the opposite side (4), and so forth. When the string becomes too short, tie it to a vertex. Then tie another meter of string to any vertex not containing three medians. Continue this process until each of the four faces looks like figure 13.10.

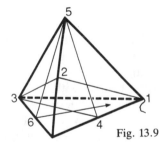

Fig. 13.9

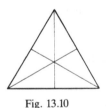

Fig. 13.10

3. Position a triangle in the large tetrahedron. Cut a string about a meter long. Then, about 7 cm from one end, tie it to the intersection of three medians at the center of a face of the tetrahedron. Do not cut off the 7-cm end because it will be needed later. Thread the larger needle to the other end of the string. Stretch the threaded needle under an edge of the tetrahedron (fig. 13.11). Pass the threaded needle through one short straw and push the straw toward the center of the first face (fig. 13.12), loop the threaded needle around the intersection of medians at the center of a second face (fig. 13.13), and pull the thread securely. This process is repeated twice to form the small triangle: add a straw to join this second face to a third face, and then attach a straw to connect the third to the first, thereby forming a triangle (fig. 13.14). Tie the string to the 7-cm end. Do not cut the strings.

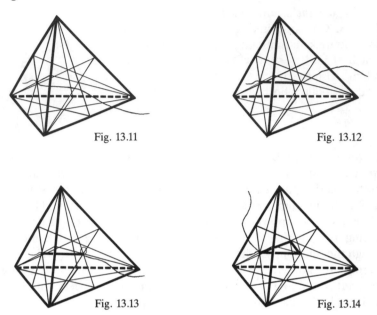

Fig. 13.11   Fig. 13.12

Fig. 13.13   Fig. 13.14

4. Complete the small tetrahedron. Add a straw connecting the small triangle to the fourth face of the large tetrahedron. Then attach another straw joining it to a vertex of the small triangle (fig. 13.15). Next pass the threaded needle through a short straw toward a vertex with only two straws (fig. 13.16). Now add the last straw and fasten it to the vertex needing a third straw (fig. 13.17). Pass the needle through a straw to get back to the 7-cm string. Tighten the thread so there is no sag in the small tetrahedron. Tie the string and cut off the extra strings.

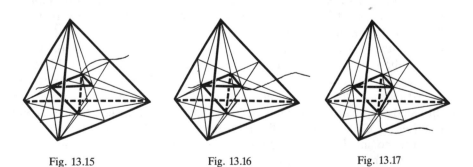

Fig. 13.15   Fig. 13.16   Fig. 13.17

Activity 4. Cube in Octahedron (for experts)

Use twelve 13-cm cocktail straws with a pinhole at their midpoints to build an octahedron. See Activity 2. Then construct a cube inside the octahedron by following the methods explained in Activity 3. The drawings below show how to begin each step.

1. Stretch the medians until each face has three medians (fig. 13.18).

2. Connect adjacent intersections of medians with twelve 5.5-cm cocktail straws (figs. 13.19a–d).

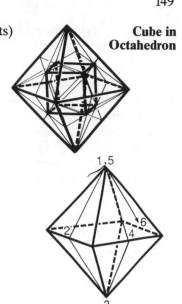

Cube in Octahedron

Fig. 13.18

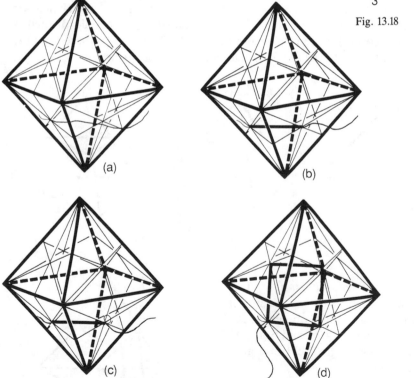

Fig. 13.19

Activity 5. Octahedron inside Tetrahedron

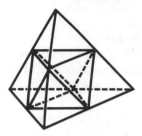

1. Prepare the straws and build a tetra-hedron. Cut twelve cocktail straws, each 6.3 cm long and put them aside until step 2. Next cut six cocktail straws to a length of 13 cm and punch a pinhole at the middle of each. Construct a tetrahedron with these six straws by following the directions in both steps of Activity 1.

2. Build a triangle on one vertex of the tetrahedron. In this step use the short straws to connect the three edges of one vertex, *A*, of the tetrahedron. Thread a rather thin needle. Pass the needle through the midpoint of one side of the tetrahedron and tie it, leaving an end of thread about 7 cm long. Pass the threaded needle through one short straw and then through a hole at the middle of a second side of the tetrahedron (fig. 13.20a). Next move the needle through a second short straw and through a central hole on another edge of vertex *A* (fig. 13.20b). Then drop the needle through a third short straw. Pull the string and tie it to the 7-cm end so there is no sag in the small triangle (fig. 13.20c). Do not cut off the string.

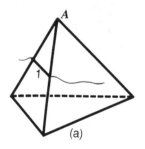 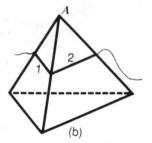 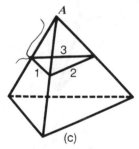

Fig. 13.20

3. Build a second triangle. Use the same string and three more short straws to construct a triangle on vertex *B*. That is, thread in straws 4, 5, and 6 (fig. 13.21). Tie the string to the 7-cm end and then cut off the ends.

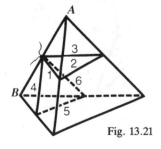

Fig. 13.21

4. Build the third and fourth triangles. Now follow steps 2 and 3 above to construct two more triangles, one on vertex *C* and one on vertex *D*. Begin by tying the string to the midpoint of the unused edge of the tetrahedron. The two drawings in figure 13.22 show these triangles numbered to fit the instructions in steps 2 and 3. After you have completed these two triangles, arrange the vertices of the octahedron neatly by gently pulling the ends of the short straws toward the exterior. You now have a beautiful octahedron inside a tetrahedron.

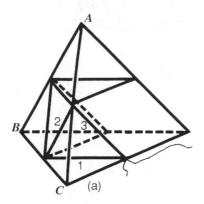 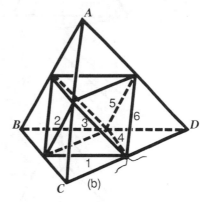

Fig. 13.22

Activity 6. Cube on Tetrahedron

Cocktail straws are not recommended for this activity because the threaded needle must be dropped through some straws six times; the needle will get caught in the other strings inside a slender straw.

1. Build the tetrahedron. Use six 19-cm straws and follow the directions in Activity 1. Also stabilize the vertices so that the tetrahedron will be strong enough to support the cube.

Cube on Tetrahedron

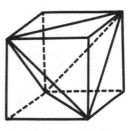

2. Build a pyramid on one face of the tetrahedron. Cut twelve 13.5-cm drinking straws. This step will use a triangular face of the tetrahedron and three short straws; that is, straws 1–6 in figure 13.23 will be used. Thread the needle and then follow the numbering in figures 13.23–13.26. Drop the threaded needle through straws 1, 4, and 5. Tie the string to hold the newly formed triangle together firmly and then drop the string through straw 1 again (fig. 13.24). Proceed to the next long straw and pull the string through straws 2, 6, 4, and back through straw 2 again (fig. 13.25). To stabilize the last three vertices of this pyramid, pull the string through straws 3, 5, 6, and 3 again (fig. 13.26). Pull the string tightly so there is no sag in the newly constructed pyramid. Then tie the string and cut off the unneeded ends.

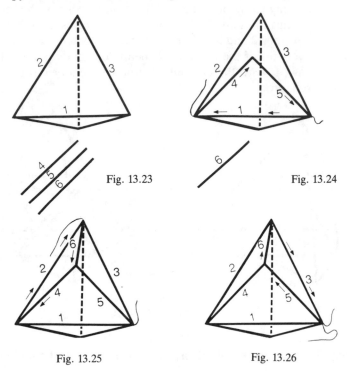

Fig. 13.23   Fig. 13.24

Fig. 13.25   Fig. 13.26

3. Build three more pyramids. Notice that the long-edged tetrahedron has three more triangular faces. Use the instructions in step 2 above to build a pyramid on each of these faces. The three drawings in figure 13.27 show these pyramids with each pyramid numbered to fit the instructions in step 2. When you have completed the last of the four pyramids, carefully pull the unsupported vertices of the cube outward.

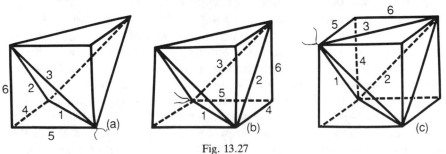

Fig. 13.27

Related Exercises

Use the exercises that follow as a guide for making your own master, choosing the bracketed word that names the polyhedron your students have constructed. See the labeling in the answer key.

In exercises 1 to 12, the student is to insert the correct number in each blank space. Exercises 7 to 12 are quite challenging, and each statement tests three separate concepts. Choose the statements (or parts of statements) that are appropriate for your students.

Exercises for Activities 1 to 6. Tell how many.

1. A/an [tetrahedron, octahedron, cube] has ____vertices, _____ edges, and _____ faces.

2. In this polyhedron, _____ edges meet at each vertex and _____ faces meet at each vertex.

3. Each edge is parallel to _____ edges, intersects _____ edges, and is skew to the other _____ edges.

4. Each edge lies in _____ faces, is parallel to _____ faces, and intersects the planes of the other _____ faces.

5. Each face is parallel to _____ of its faces and intersects the planes of the other _____ faces.

6. This polyhedron has _____ dihedral angles and _____ polyhedral angles.

Exercises for Activities 3 to 6. Tell how many.

7. Each edge of the outer [tetrahedron, octahedron, cube] is parallel to _____ edges of the inner [tetrahedron, octahedron, cube], intersects _____ edges, and is skew to the other _____ edges.

8. Each edge of the inner polyhedron is parallel to _____ edges of the outer polyhedron, intersects _____ edges, and is skew to the other _____ edges.

9. Consider the plane containing one face of the outer polyhedron. This plane contains _____ faces of the inner polyhedron, is parallel to _____ faces, and intersects the planes of the other _____ faces.

10. Consider the plane containing one face of the inner polyhedron. This plane contains _____ faces of the outer polyhedron, is parallel to _____faces, and intersects the planes of the other _____ faces.

11. Consider the line determined by an edge of the outer polyhedron. This line is contained in the plane of _____ faces of the inner polyhedron, is parallel to _____ faces, and intersects the planes of the other ____ faces.

12. Consider the line determined by an edge of the inner polyhedron. This line is contained in _____ faces of the outer polyhedron, is parallel to _____ faces, and intersects the planes of the other _____ faces.

The constructed models supply further instances of geometric ideas. Ask students to find and describe examples of the following concepts:

- Lines of symmetry for a face
- Lines of symmetry for a polyhedron
- Two parallel lines determining a plane
- Two intersecting lines determining a plane
- Three noncollinear points determining a plane
- Two planes intersecting in a line
- Noncoplanar points
- Duals

Point out that in Activity 6 the cube would collapse without the support of the tetrahedron because three points determine a plane, whereas the cube has four vertices in each of its faces.

Answers

Tetrahedron (Activities 1, 3, 5, 6):
1. 4, 6, 4   **2.** 3, 3   **3.** 0, 4, 1   **4.** 2, 0, 2   **5.** 0, 3   **6.** 6, 4

Octahedron (Activities 2, 4, 5):
1. 6, 12, 8   **2.** 4, 4   **3.** 1, 6, 4   **4.** 2, 2, 4   **5.** 1, 6   **6.** 12, 6

Cube (Activity 6):
1. 8, 12, 6   **2.** 3, 3   **3.** 3, 4, 4   **4.** 2, 2, 2   **5.** 1, 4   **6.** 12, 8

Tetrahedron in Tetrahedron (Activity 3):
7. 1, 0, 5   **8.** 1, 0, 5   **9.** 0, 1, 3   **10.** 0, 1, 3   **11.** 0, 2, 2   **12.** 0, 2, 2

Cube in Octahedron (Activity 4):
7. 0, 0, 12   **8.** 0, 0, 12   **9.** 0, 0, 12   **10.** 0, 0, 12   **11.** 0, 2, 4   **12.** 0, 0, 8

Octahedron in Tetrahedron (Activity 5):
7. 2, 4, 6   **8.** 1, 2, 3   **9.** 1, 1, 6   **10.** 1, 0, 3   **11.** 2, 2, 4   **12.** 1, 1, 2

Tetrahedron in Cube (Activity 6):
7. 0, 3, 3   **8.** 0, 6, 6   **9.** 0, 0, 4   **10.** 0, 0, 6   **11.** 0, 0, 4   **12.** 1, 1, 4

REFERENCE

Pohl, Victoria. *How to Enrich Geometry Using String Designs.* Reston, Va.: National Council of Teachers of Mathematics, 1986.

14

Conic Sections:
An Exciting Enrichment Topic

Roselyn Teukolsky

It is a joy and a challenge to teach students whose idea of fun is exploring the golden ratio. For them, happiness is the creation of an elegant original proof. In a class of such students a teacher can serve up not only the staples of Euclidean and solid geometry but also a feast of gourmet fare. The possibilities for enrichment are limitless.

Conic sections are traditionally given short shrift in geometry curricula. It is especially regrettable to deny them to gifted students, who have the ability and insight to appreciate their geometric magic. I propose that conic sections be taught as the exciting grand finale of an enriched geometry course. Beautiful and fascinating in their many aspects, they offer a rare opportunity to blend analytic and solid geometry, locus, similar triangles, circles and spheres, and so on, into a potpourri of unusual and unexpected results. Through conics, geometry is shown to be part of a consistent mathematical whole rather than some disembodied, esoteric discipline.

What follows is a unit for gifted students in which the conics are defined and generated in a variety of ways. The unifying theme, then, is to show that the definitions are all mathematically equivalent. As the unit progresses, the aim is to amaze the students: How could curves generated in such different ways each time lead to the same conic sections? In the end, the students will be enraptured by the web of interconnections.

Introduction

Legend has it that conic sections originated around 430 B.C. as a result of plague in Athens. Through the Oracle of Delos, Zeus announced to the suffering citizens that to end the plague they would have to construct an altar twice the size of the existing cubical one to Apollo. All straightedge and compass attempts to double the cube failed. About 340 B.C. Menaech-

I would like to thank Avery Solomon for sparking my interest in conic sections, Bill Halton for invaluable ideas and discussions, and Saul Teukolsky for assistance with the manuscript.

155

mus found two solutions using conics. The fate of the altar is not of interest here. The question is, How were the earliest conics described? The writings of Menaechmus have been lost. However, according to Geminus the ancients used only right cones in their definition of the conics. Of these, they distinguished three types—those with acute, right, or obtuse vertex angles (Heath 1921, p. 111). The three conic sections were then produced by a cutting plane perpendicular to the rulings of the cone (fig. 14.1).

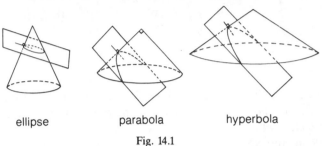

ellipse   parabola   hyperbola

Fig. 14.1

Apollonius of Perga (262–200 B.C.) gave the authoritative treatise (eight volumes' worth) on conics. His great advance was to generate all the conics from a single, oblique, double cone, simply by varying the tilt of the cutting plane (fig. 14.2). Apollonius is also credited with coining the names *parabola, ellipse,* and *hyperbola.*

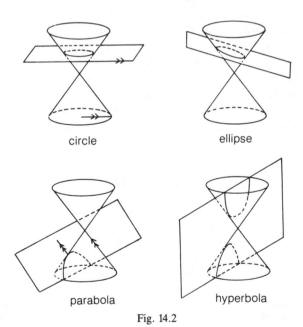

circle   ellipse

parabola   hyperbola

Fig. 14.2

If the cutting plane passes through the vertex *V* of the cone, a *degenerate conic* is obtained. A degenerate conic may be just one point, a line, or two intersecting lines. Here we consider only the nondegenerate conics.

Sections of a Cone

Let us look at the nature of the nondegenerate conics more closely (Bridge 1831, p. 10). In each example we shall consider a right circular cone cut by the following planes (see figs. 14.3–14.5):

- *CPD,* which is any plane parallel to a base of the cone. Its intersection with the cone is a circle.
- *BEG,* which is perpendicular to the base. Its intersection with the cone is △*BEG.* Note that plane *BEG* intersects both circular bases in their diameters.
- Cutting plane *APO,* which is perpendicular to plane *BEG.* If *APO* is parallel to a plane containing $\overleftrightarrow{BE}$ and tangent to the cone at $\overleftrightarrow{BE}$, then the intersection of *APO* and the cone is a parabola (fig. 14.3). If *APO* passes through both sides $\overleftrightarrow{BG}$ and $\overleftrightarrow{BE}$ of the cone, the intersection is an ellipse (fig. 14.4). If *APO* is not parallel to a tangent plane containing $\overleftrightarrow{BE}$ and does not intersect $\overline{BE}$, the intersection is a hyperbola (fig. 14.5).

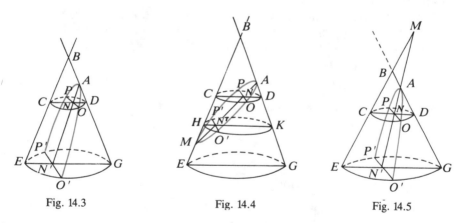

Fig. 14.3   Fig. 14.4   Fig. 14.5

Note that in each case the intersection of planes *APO* and *BEG* is $\overleftrightarrow{ANN'}$, which is an axis of symmetry of the conic. Point *A* is a vertex. By plane perpendicularity $\overline{PNO} \perp$ plane $\overline{BEG}$, $\overline{PNO} \perp \overleftrightarrow{ANN'}$, and $\overline{PNO} \perp \overline{CND}$.

Now refer to figure 14.3 and the parabola. Again by plane perpendicularity $\overline{P'N'O'}$ is perpendicular to plane *BEG*, $\overleftrightarrow{ANN'}$, and $\overline{EN'G}$. Using chord properties in a circle, we see that

$$PN = NO, \quad P'N' = N'O'.$$

Thus, by the two-chord power theorem,

$$CN \cdot ND = PN^2.$$

Therefore,

$$ND = \frac{PN^2}{CN}. \tag{1}$$

Similarly,

$$N'G = \frac{P'N'^2}{EN'}. \tag{2}$$

$CNN'E$ is a parallelogram, since both pairs of opposite sides are parallel. Therefore, $CN = EN'$. Substituting in (2) gives

$$N'G = \frac{P'N'^2}{CN}. \tag{3}$$

Using $\triangle AND \sim \triangle AN'G$ and substituting from (1) and (3), we have

$$\frac{AN}{AN'} = \frac{ND}{N'G} = \frac{PN^2/CN}{P'N'^2/CN} = \frac{PN^2}{P'N'^2}.$$

Thus a parabola is characterized by this property: If its vertex is A, P is any other point on the curve, and a segment $\overrightarrow{PN} \perp \overleftrightarrow{AN}$, then AN is directly proportional to PN^2.

Refer to figure 14.4 and the ellipse. Point M on $\overleftrightarrow{BE}$ is a second vertex of the ellipse. Let another plane $HP'K$ be taken parallel to the base such that it intersects plane APO as shown. Then $\overrightarrow{P'N'O'} \perp \overline{ANN'}$ and $\overrightarrow{P'N'O'} \perp \overline{HK}$. As for the parabola,

$$NC \cdot ND = PN^2 \text{ and } HN' \cdot N'K = P'N'^2. \tag{4}$$

Since $\triangle AND \sim \triangle AN'K$ and $\triangle MNC \sim \triangle MN'H$,

$$\frac{AN}{AN'} = \frac{ND}{N'K} \text{ and } \frac{NM}{N'M} = \frac{NC}{N'H}.$$

$$\therefore \frac{AN}{AN'} \cdot \frac{NM}{N'M} = \frac{ND}{N'K} \cdot \frac{NC}{N'H}$$

Substituting from (4) yields

$$\frac{AN \cdot NM}{AN' \cdot N'M} = \frac{PN^2}{P'N'^2}.$$

Thus an ellipse is characterized by this property: If points A and M are the vertices of its longer (major) axis and P is any point on the curve with $\overrightarrow{PN} \perp \overleftrightarrow{AN}$, then $AN \cdot NM$ is directly proportional to PN^2.

Refer to figure 14.5 and the hyperbola. Let plane APO meet $\overrightarrow{EB}$ in M.

(This represents the point at which plane APO cuts the upper nappe of the cone to form the second branch of the hyperbola.) As before, since

$$\triangle AND \sim \triangle AN'G \text{ and } \triangle MNC \sim \triangle MN'E,$$

$$\frac{AN}{AN'} = \frac{ND}{N'G} \text{ and } \frac{NM}{N'M} = \frac{NC}{N'E}.$$

$$\therefore \frac{AN \cdot NM}{AN' \cdot N'M} = \frac{ND \cdot NC}{N'G \cdot N'E} = \frac{PN^2}{P'N'^2}$$

Thus a hyperbola is characterized by this property: If points A and M are the vertices and P any point on the curve with $\overleftrightarrow{PN} \perp \overleftrightarrow{AN}$, then $AN \cdot NM$ is directly proportional to PN^2.

Focal Distances

Now consider two fixed points, F and F'. We wish to examine the locus of all points P in a given plane for which $FP + F'P$ is a constant. To do this, fasten the ends of a piece of string to two pins fixed on the chalkboard at F and F' and trace a curve on the board with chalk pressed against the string, keeping it taut (fig. 14.6). Let the distance FF' be $2c$ and the length of the string be $2a$. Thus for all points P of the locus, $FP + F'P = 2a$. Furthermore, if P_1 and P_2 are vertices of the curve, then $P_1P_2 = 2a$; for when the position of the chalk is at P_2, the length of the string is $2a = FF' + 2F'P_2 = FF' + F'P_2 + FP_1 = P_1P_2$.

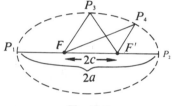

Fig. 14.6

The curve obtained certainly looks like an ellipse. But do the points on it exhibit the proportionality property proved above for the elliptical section of the cone? One way to investigate this is to derive an equation for the curve. Choose the axes as shown (fig. 14.7), and let $P(x,y)$ be any point on the curve. Then $FP + F'P = 2a$; therefore,

$$\sqrt{(x + c)^2 + (y - 0)^2} + \sqrt{(x - c)^2 + (y - 0)^2} = 2a.$$

Taking the second radical over to the right-hand side and squaring both sides gives

$$(x + c)^2 + y^2 = 4a^2 - 4a\sqrt{(x - c)^2 + y^2} + (x - c)^2 + y^2.$$

Simplifying, we have

$$a^2 - cx = a\sqrt{(x - c)^2 + y^2}.$$

Squaring both sides and simplifying again, we have

$$(a^2 - c^2)x^2 + a^2y^2 = (a^2 - c^2)a^2. \tag{5}$$

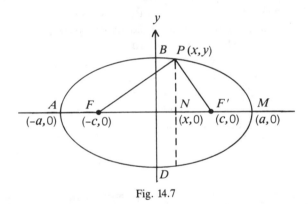

Fig. 14.7

At this point recall what it is we are trying to discover: Does this curve have the proportionality property $AN \cdot NM = k \cdot PN^2$, where k is some constant? From equation (5),

$$(a^2 - c^2)(a^2 - x^2) = a^2y^2.$$

$$\therefore (a + x)(a - x) = \frac{a^2}{a^2 - c^2}y^2.$$

Since $AN = a + x$, $NM = a - x$, $PN = y$, and $a^2/(a^2 - c^2)$ is constant for all positions of P, we have shown that $AN \cdot NM = k \cdot PN^2$, and the curve is an ellipse. (Note that the proportionality property *uniquely* characterizes the ellipse; for given a curve satisfying this property, one can choose axes as in figure 14.7 and derive equation (5). Since the equation is of second degree, its roots determine only two loci $y = y(x)$. But these are simply the upper and lower branches of the ellipse. Similarly, the parabola and hyperbola are uniquely characterized by their proportionality properties.)

Several interesting features of the ellipse may be noted here. In equation (5) above, set $b^2 = a^2 - c^2$. (Since $a > c$, this is permissible.) Then $b^2x^2 + a^2y^2 = b^2a^2$, and dividing by a^2b^2 gives a standard form of the equation:

$$\frac{x^2}{a^2} + \frac{y^2}{b^2} = 1$$

Notice that $y = 0$ yields $x = \pm a$, and $x = 0$ yields $y = \pm b$. Thus the coordinates of points B and D are $(0,b)$ and $(0, -b)$ respectively. $\overline{AM}$ is the *major* and $\overline{BD}$ the *minor axis* of the ellipse. F and F' are called the *foci* of the ellipse, and $e = c/a$ is called the *eccentricity* of the ellipse. Since $c < a, e < 1$. As $b \to a$, $c \to 0$; therefore, $e \to 0$, forming a "fatter" ellipse. At $c = 0$ and $e = 0$ a circle is obtained with both foci at the same point, the center of the circle.

A similar analysis can be given for the hyperbola. Again consider two fixed points F and F'. We wish to find the locus of points P in a given plane for which $|FP - F'P|$ is a constant. The mechanical construction is quite tricky, but it can be done with three pairs of nimble hands (Besant 1895).

Take a rod and a piece of string whose length is less than that of the rod. Fix one end of the string to one end of the rod L and the other end of the string to the board at F. Fix the other end of the rod to the board at F'. Thus the rod is movable in the plane of the board (fig. 14.8). Press a piece of chalk against the string, keeping the string taut with a part of it, $\overline{CL}$, in contact with the rod. The chalk will trace out a curve as shown in figure 14.8b. Note that for any position C of the chalk,

$$CF' - CF = LC + CF' - (CF + LC)$$
$$= \text{length of rod} - \text{length of string}$$
$$= \text{a constant}$$
$$= 2a, \text{ say, for some } a > 0.$$

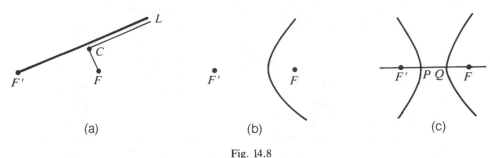

(a)   (b)   (c)

Fig. 14.8

Now perform the identical construction (using the same F and F') with a piece of string that exceeds the length of the rod by $2a$.

These two constructions give two branches of what appears to be a hyperbola. Note that for the point P shown in figure 14.8c,

$$PF - PF' = PF - QF = PQ = 2a.$$

Do the points on this curve have the proportionality property proved for the hyperbola on the cone? Showing that they do is an excellent exercise for students to complete on their own, since the algebraic manipulation is analogous to that required for the ellipse.

They should again let $FF' = 2c$ and choose axes as shown (fig. 14.9).

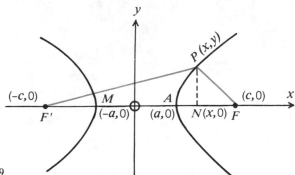

Fig. 14.9

Then using the fact that $|PF - PF'| = 2a$, they can derive the equation

$$x^2(c^2 - a^2) - a^2(c^2 - a^2) = a^2y^2. \tag{6}$$

They next must show that

$$AN \cdot NM = k \cdot PN^2.$$

Thus the curve is a hyperbola.

Now, defining $c^2 - a^2 = b^2$ in equation (6) will yield the standard form of the equation:

$$\frac{x^2}{a^2} - \frac{y^2}{b^2} = 1$$

Once more, the eccentricity $e = c/a$. Note that this is always greater than 1.

We now have an alternative definition for the ellipse and hyperbola: *Let F and F' be two fixed points called foci. The set of points P in a given plane for which FP + F'P is a constant is an ellipse. The set of points for which |FP − F'P| is a constant less than FF' is a hyperbola.*

Focus-Directrix

Here is a different locus to consider. A line l and point F not on l determine a plane; a point P moves in this plane in such a way that the ratio of its distance from F to its distance from l is always the same. What kind of curve does P trace out?

Denote the distance from the fixed point F as p and that from the line as d. There are three cases to consider: $p/d = 1$, $p/d < 1$, and $p/d > 1$. Each student should therefore plot three different graphs. In doing so, they should recall that in a given plane the locus of points a given distance from a line is a pair of parallel lines one on either side of the line, whereas the locus of points a given distance from a point is a circle. The intersection of these loci satisfy both conditions. The graphs obtained look something like those in figure 14.10.

It appears that we have another definition for the parabola, ellipse, and hyperbola. Can we show that this "point-line" definition is consistent with what has gone before?

We shall start with the parabola. Using a line l, a fixed point F, and the new definition, we derive the equation of the curve. Choose axes as shown in figure 14.11. The y-axis is parallel to l and midway between F and l. Let $P(x,y)$ be any point on the curve. Then

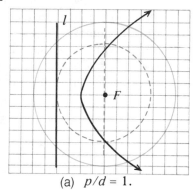

(a) $p/d = 1$.

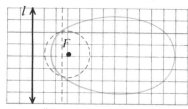

(b) $p/d = 3/4 < 1$.

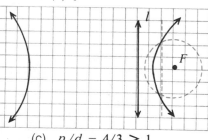

(c) $p/d = 4/3 > 1$.

Fig. 14.10

$$PM = PF.$$
$$\therefore x + a = \sqrt{(x - a)^2 + y^2}$$
$$\therefore x^2 + 2ax + a^2 = (x - a)^2 + y^2$$
$$\therefore y^2 = 4ax$$

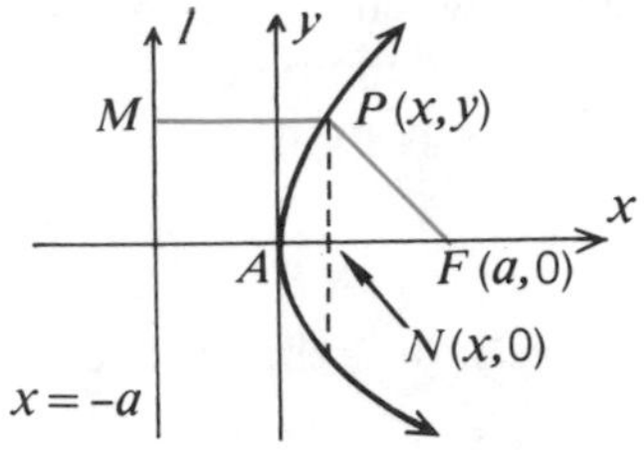

Fig. 14.11

Since $AN = x$, $PN = y$, and $4a$ is a constant, AN is proportional to PN^2, which satisfies the condition for the parabola on the cone.

For the curve that appears to be an ellipse: Can we show that it *is* an ellipse by proving that it possesses the "sum of focal distances" property stated in the previous section? Notice from the graph that the curve has an axis of symmetry through F that is perpendicular to l. This intersects l at D and the curve at A, say (fig. 14.12). Take point F' and line l' as shown, such that $AF = BF'$ and $AD = BC$. This ensures that for any point P on the curve, p/d is the same with respect to F' and l'. Thus,

$$\frac{PF}{PM} = k \quad \text{and} \quad \frac{PF'}{PN} = k.$$

Therefore,

$$PF + PF' = kPM + kPN = kMN, \text{ a constant.}$$

Therefore the curve is an ellipse with foci at F and F'.

Furthermore, using the quantities a and c previously defined, we can fill in lengths as shown (fig. 14.13). With respect to point A:

$$AF/AD = AF'/AC$$
$$\therefore \frac{p}{d} = \frac{p + 2c}{2a + d}$$
$$\therefore 2ap + pd = pd + 2cd$$
$$\therefore p/d = c/a$$
$$= e, \text{ the eccentricity of the ellipse}$$

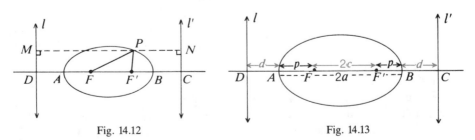

Fig. 14.12   Fig. 14.13

At this stage the students will know that a similar analysis awaits the hyperbola look-alike obtained graphically when $p/d > 1$. Perhaps they can try it for homework.

They should obtain a diagram as for the ellipse, with distances marked as shown in figure 14.14:

$$\frac{PF}{PM} = k \quad \text{and} \quad \frac{PF'}{PN} = k$$

Thus

$$|PF' - PF| = kPN - kPM = kMN, \text{ a constant.}$$

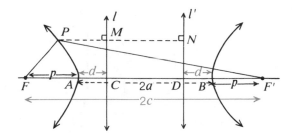

Fig. 14.14

Therefore, the curve is a hyperbola with foci at F and F'. Also,

$$AF/AC = AF'/AD$$
$$\therefore \frac{p}{d} = \frac{2c - p}{2a - d}$$
$$\therefore p/d = c/a = e, \text{ the eccentricity.}$$

We now have a new definition for a conic: *Given a line l, called the directrix, and a point F not on l, called the focus, a conic section is the locus of all points P for which the ratio*

$$\frac{\textit{distance from P to F}}{\textit{distance from P to l}}$$

is a constant. This constant is called the eccentricity e of the conic, which is an ellipse if $0 < e < 1$, a parabola if $e = 1$, and a hyperbola if $e > 1$.

Waxed Paper

For a different kind of lesson, give each student three pieces of waxed paper, about 12 by 9 inches. On the first piece they should rule a line and take a point not on the line. They should then fold this point onto the line at many different points—as many as possible in about five minutes. Each fold is etched into the wax as a visible line, and soon a picture emerges.

On the second piece of waxed paper they should make a circle about four inches in diameter and take a point, other than the center, in the interior of the circle. This point should be folded onto the circle at as many different points as possible in the time allowed.

On the third paper they can make another circle and take a point outside it. As before, the point is folded onto the circle.

The effect is dramatic when the pieces of waxed paper are viewed on a dark surface: like magic, our old friends the parabola, ellipse, and hyperbola seem to have reappeared, beautifully framed in envelopes of tangents (fig. 14.15).

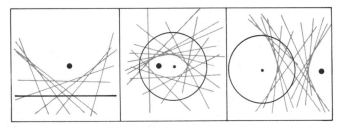

Fig. 14.15

A rigorous proof that the wax parabola is in fact the same parabola we defined previously requires some knowledge of the tangent properties of conics, which is beyond the scope of this unit. We can, however, make a convincing intuitive argument. A careful examination of the points on the curve (fig. 14.16) should reveal that the curve is the locus of points X that lie both on t, the perpendicular bisector of $\overline{FP}$ where P is the "folding point" on d and t is the fold, and on the perpendicular erected at P. When F is folded onto Q, which is on the curve's axis, point A, the vertex of the curve, is obtained. Note that t is the tangent to the curve at X. Since X is on the perpendicular bisector of $\overline{FP}, X$ is equidistant from P from F. Therefore the curve is a parabola with focus F and directrix d.

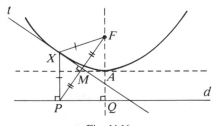

Fig. 14.16

Now examine the points on the wax ellipse (fig. 14.17). F is the fixed point that was folded onto different points of the circle. Notice that the curve is the locus of points X that lie both on t, the perpendicular bisector of $\overline{FP}$, where P is the folding point on the circle and t is the fold, and on segment $\overline{OP}$, where O is the center of the circle. Note that the fold t is the tangent to the curve at X.

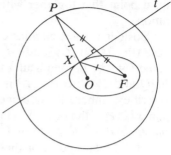

Fig. 14.17

To show that the curve is an ellipse:
Draw $\overline{XF}$. Since X is on the perpendicular bisector of $\overline{PF}$,

$$XO + XF = XO + XP$$
$$= OP$$
$$= \text{the radius of the circle, a constant.}$$

Therefore the curve is an ellipse with foci at O and F.

The wax hyperbola can be similarly analyzed (fig. 14.18). O is the center of the circle. F is the point in the exterior of the circle. P is the folding point. Point X is the intersection of fold t (the perpendicular bisector of $\overline{PF}$), and $\overrightarrow{OP}$. Notice that t is a tangent to the curve at X.

$$XF - XO = XP - XO$$
$$= OP$$
$$= \text{the radius of the circle,}$$
$$\text{a constant.}$$

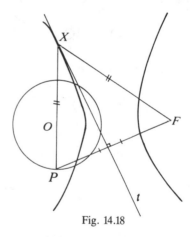

Fig. 14.18

Therefore the curve is a hyperbola with foci at O and F.

The Cone Revisited

Are our original definitions of the parabola, ellipse, and hyperbola as sections of a cone consistent with the focal-distances and focus-directrix definitions (Besant 1895)? And if they are, where are the foci and directrices with respect to the cone? Our aim in this section is to find (1) a fixed point and line for which p/d is constant for all points on the curve and (2) another fixed point that together with the first will satisfy the focal-distances definition.

Consider a right circular cone cut by plane OVQ perpendicular to the base (fig. 14.19). $\overline{OV}$ and $\overline{OQ}$ are generating lines of the cone. Let cutting plane UAP be perpendicular to plane OVQ. $\overleftrightarrow{AU}$ is the intersection of planes VOQ and UAP; curve AP is the intersection of plane UAP and the cone. Inscribe a sphere in the cone, touching the cone in the circle EF and the plane UAP in the point S. The intersection of the plane of the circle EF and plane UAP is $\overleftrightarrow{XK}$. Note that $\overleftrightarrow{XK}$ is perpendicular to plane VOQ.

Let P be any point of the curve. Draw $\overline{OP}$, which cuts circle EF in R, say. Draw $\overline{SP}$. Through P let QP be the circular section parallel to a base, and let it cut plane UAP in $\overleftrightarrow{PN}$. $\overleftrightarrow{PN}$ is therefore perpendicular to $\overleftrightarrow{AN}$ and parallel to $\overleftrightarrow{XK}$. $XKPN$ is a parallelogram.

Since $\overline{SP}$, $\overline{PR}$, $\overline{AE}$, and $\overline{AS}$ are tangents to the sphere,

$$SP = RP = EQ, \tag{7}$$
$$AE = AS. \tag{8}$$

In plane OVQ, $\triangle QNA \sim \triangle EXA$. Therefore,

$$\frac{AN}{AX} = \frac{AQ}{AE}$$

$$\therefore \frac{AN + AX}{AX} = \frac{AQ + AE}{AE}$$

$$\therefore \frac{NX}{AX} = \frac{EQ}{AE}$$

$$\therefore \frac{EQ}{NX} = \frac{AE}{AX}$$

$$= \frac{AS}{AX}. \tag{9}$$

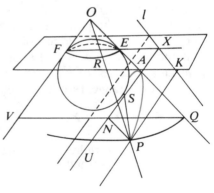

Fig. 14.19

But since $PK = NX$, (7) and (9) give

$$\frac{SP}{PK} = \frac{AS}{AX}. \tag{10}$$

AS/AX is a constant, independent of the choice of P. The curve AP is therefore an ellipse, parabola, or hyperbola, depending on whether AS is less than, equal to, or greater than AX. In all cases S is a focus, and l, the intersection of the plane of the curve with the plane of contact of the sphere and cone, is a corresponding directrix. Let us now consider each case in turn.

Case 1: The Parabola

Plane UAP is parallel to $\overleftrightarrow{OV}$. Then

$$\angle AXE = \angle OFE = \angle OEF = \angle AEX,$$

so $\triangle AEX$ is isosceles. Therefore,

$$AS = AE = AX.$$

Then in (10) above, $SP/PK = 1$, and the curve is a parabola.

Case 2: The Ellipse

Plane UAP cuts $\overrightarrow{OQ}$ and $\overrightarrow{OV}$ in A and A' as shown in figure 14.20. Compare the orientation of plane UAP in figures 14.19 and 14.20. In figure 14.20 the plane has been "lifted up." Thus $\angle AEX > \angle FXA$.

$$\therefore AE < AX$$
$$\therefore AS < AX$$

In (10) above,

$$SP/PK = AS/AX < 1,$$

and therefore the curve is an ellipse.

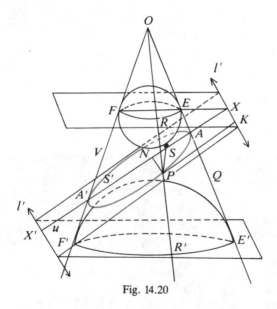

Fig. 14.20

In this case another sphere can be inscribed in the cone as shown, touching the cone along the circle $E'F'$ and touching the plane UAP at S'. It can be shown as before that S' is a focus and that the corresponding directrix is l', the intersection of plane UAP and the plane containing circle $E'F'$. Is this placement of the foci consistent with the focal-distances definition? In other words, can it be shown that $PS + PS'$ is a constant? Using congruent tangent segments to a sphere from an external point, we can show that

$$PS + PS' = PR + PR'$$
$$= RR'.$$

This is constant for any point P on the ellipse.

Case 3: The Hyperbola

Here plane UAP intersects both nappes of the cone, forming two branches, as shown in figure 14.21. This time,

$$\angle AEX < \angle AXF$$
$$\therefore AE > AX$$
$$\therefore AS > AX.$$

In (10) above,

$$\frac{SP}{PK} = \frac{AS}{AX} > 1,$$

and therefore the curve is a hyperbola. In this case another sphere can be inscribed in the other nappe of the cone, touching the cone along the circle $E'F'$ and plane $UA'P'$ in S'. Taking point P' on this branch, it follows that

$$SP'/P'K' = SA/AX.$$

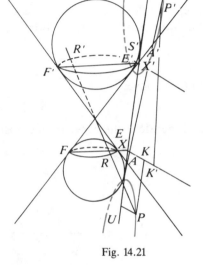

Fig. 14.21

Thus S' is the other focus of the hyperbola, and the corresponding directrix is the intersection of the cutting plane and the plane containing circle $E'F'$.

Again the question can be posed, Are these placements of the foci consistent with the focal-distances definition? In other words, can it be shown that $PS - PS'$ is a constant? Using congruent tangent segments as for the ellipse, we have

$$PS' - PS = PR' - PR = RR',$$

which is constant for any point P on the hyperbola.

This elegant "spheres" method of determining the focus and directrix was first published by Pierce Morton of Trinity College in the first volume of the Cambridge *Philosophical Transactions* (Besant 1895).

Some interesting challenges to the students: Try to provide arguments that (1) parallel parabolas and (2) parallel hyperbolas on a given cone have the same eccentricity. Also, try to prove that for a given cone, parallel planes cut sections with equal eccentricity.

The Cylinder

One of the fascinations of the conics is that they can be generated in a variety of seemingly unconnected ways that turn out to be mathematically equivalent. The students who understand and appreciate this fact will enjoy trying their hands at the following problem and also the one in the next section.

- Prove that if a right circular cylinder is cut by a plane not parallel to its base, then the intersection is an ellipse. Discuss the positions of the foci and directrices of this ellipse with respect to the cylinder.

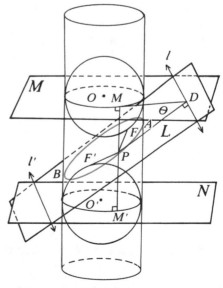

Here is a solution (fig. 14.22). Let $\mathcal{L}$ be the cutting plane and curve APB the intersection of $\mathcal{L}$ and the cylinder. Consider spheres tangent to the cylinder and to $\mathcal{L}$, with the circles (centers O and O') being the circles of tangency and F and F' being the points of tangency of the upper and lower

Fig. 14.22

spheres respectively. Let planes *M* and *N* contain circles *O* and *O'*, respectively. Let the intersection of *∠* and *M* be *l* and of *∠* and *N* be *l'*. *P* is any point on curve *APB*. Through *P* draw the line on the cylinder perpendicular to the base, and let this intersect circles *O* and *O'* in *M* and *M'*, respectively. Let the measure of the dihedral angle formed by *∠* and *M* be θ.

To show that the intersection of *∠* and the cylinder is an ellipse with foci at *F* and *F'*, we must prove that *PF* + *PF'* is constant:

$$PF + PF' = PM + PM'$$
$$= MM', \text{ a constant distance.}$$

To show that *l* and *l'* are directrices, we must prove that for each of them *p/d* is a constant. Draw $\overline{PD} \perp l$. Draw $\overline{MD}$. Since $\overline{PM} \perp M$, △*PMD* is a right triangle. Therefore,

$$PF/PD = PM/PD = \sin θ.$$

This is a constant, since the dihedral angle θ is constant for any position of *P*, and therefore *l* is a directrix. It can similarly be shown that *l'* is a directrix.

Projection of a Circle

Here is the second problem for students to attempt themselves. Show that if a circle is projected onto a plane that is neither parallel nor perpendicular to the plane of the circle, then the projection is an ellipse (Besant 1895).

Here is a solution (fig. 14.23). Let *A'B'C'* be the projection of circle *ABC*. Let $\overline{PQ}$ be a chord parallel to the plane of projection. Then its projection $\overline{P'Q'}$ is congruent to $\overline{PQ}$. Let diameter $\overline{ANC} \perp \overline{PQ}$ meet the plane of projection in *F*. The projection of $\overline{ACF}$ is $\overline{A'C'F}$. Note that $\overline{A'C'}$ bisects $\overline{P'Q'}$ at right angles in the point *N'*. We must show that *A'N'* · *C'N'* is proportional to *P'N'²*.

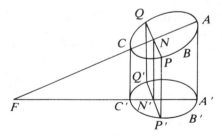

Fig. 14.23

By the basic proportionality theorem, in △*AFA'*,

$$\frac{A'N'}{AN} = \frac{A'F}{AF}.$$

By the invariance of ratios under projection,

$$\frac{C'N'}{CN} = \frac{A'F}{AF}, \quad \therefore \quad \frac{A'N}{AN} \cdot \frac{C'N'}{CN} = \frac{A'F^2}{AF^2}.$$

But in the circle

$$AN \cdot CN = PN^2 = P'N'^2, \quad \therefore \quad \frac{A'N' \cdot C'N'}{P'N'^2} = \frac{A'F^2}{AF^2}.$$

Since $A'F^2/AF^2$ is a constant independent of P,

$$A'N' \cdot C'N' = kP'N'^2,$$

which is the characteristic property of an ellipse.

Treasure Chest

This article provides just a glimpse into the treasure chest of conic sections. The definitions lead to a wealth of properties. The latus rectum and its long list of interesting characteristics have not been touched on at all. Tangents, chords, asymptotes, and diameters have all made brief appearances but have hardly been examined. In addition, several odd and unrelated facts, like small gems, lie waiting to be discovered:

- If an ellipse has its foci at the endpoints of the diameter of a circle, then the area of the ellipse equals that of the circle when the ratio of major to minor axis is the golden ratio.
- Before Pascal was sixteen he proved that if six points are taken on a conic and joined, a "mystic hexagram" is formed with this property: The three pairs of opposite sides intersect in three collinear points (Bell 1937, p. 84).
- Brianchon's theorem: If a, b, c, d, e, and f are tangents to a conic section, the lines joining the pairs of intersections a with b and d with e, b with c and e with f, and c with d and f with a are concurrent (Bell 1937, p. 238).

The practical applications of the conics add further sparkle (Whitt 1981, pp. 15, 61, 70):

- Diocles, in his book *On Burning Mirrors* (second century B.C.), came up with the following: If a victim was to be sacrificed in front of a large crowd, he or she could be placed at the focus of a parabolic mirror, which would ignite a visible burning spot on the body. It is not known whether this idea was ever implemented; however, the word *focus* is Latin for fireplace.

- Parabolic reflectors can collect sound waves in a way that enables distant conversations to be heard. Such devices have successfully been used for recording bird songs and presumably also in the world of espionage.

- The "focal distances" property of an ellipse can be used to create a "whispering gallery." In a room with an elliptical ceiling, if one speaks softly at one focus, the sound is reflected to the other focus, where it can be clearly heard. The story has been told that some churches in Europe use whispering galleries as confessionals by placing the priest and the penitent at the foci.

- The British physicist Ernest Rutherford used the hyperbolic orbits of scattered α-particles to discover the nucleus of the atom. He bombarded a thin piece of gold foil with α-particles and found that some particles were scattered through large angles in hyperbolic orbits. If, as was generally believed, the positive charge of the atom were spread out rather than being concentrated in a nucleus, such large-angle scattering would not have been observed.

The practical applications of the conics make them a relevant and fascinating topic to teach. The wealth of mathematics concealed in their graceful curves makes them a must to teach to gifted students. As H. J. S. Smith, in an 1873 presidential address to the British Association for the Advancement of Science, said (Whitt 1981, p. 1):

> If we may use the great names of Kepler and Newton to signify stages in the progress of human discovery, it is not too much to say that without the treatises of the Greek geometers on the conic sections there could have been no Kepler, without Kepler no Newton, and without Newton, no science in the modern sense of the term.

REFERENCES

Bell, E. T. *Men of Mathematics*. New York: Simon & Schuster, 1937.

Besant, W. H. *Conic Sections Treated Geometrically*. 9th ed. London: George Bell, 1895.

Bridge, G. *A Treatise on the Construction, Properties, and Analogies of the Three Conic Sections*. New Haven, Conn.: Durrie & Peck, 1831.

Heath, Thomas L. *A History of Greek Mathematics*. Vol. 2. Oxford: Clarendon Press, 1921.

Whitt, L. *The Standup Conic*. College Station, Tex.: Texas A & M University Press, 1981.

15

Probability in High School Geometry

Ernest Woodward

Larry Hoehn

THE many uses of probability in today's world makes us increasingly aware of the need to include it in our mathematics curriculum. Most students, however, complete high school with only minimal exposure to probability. As introduced in elementary and middle school, probability usually involves only elementary concepts and depends heavily on the use of dice, coins, and playing cards. Although some high school students do take a formal course in probability or probability and statistics, many schools do not offer such courses, and enrollment is usually low even when it is offered.

An ideal way for students to learn more about this important topic is for probability concepts to be taught or reinforced in other high school mathematics courses. The geometry course is an appropriate place to do this for several reasons:

1. Geometrical probability problems are quite easy to pose.
2. Geometrical probability problems are often inherently interesting and can provide motivation.
3. Students get a chance to apply previously learned geometric concepts in different and surprising ways.
4. Most importantly, students will have a better understanding of probability as a result of seeing important concepts applied in the context of geometry.

Since probability is not usually taught in high school geometry, how can this be accomplished? One answer is the introduction of a separate probability unit in the present course. A better alternative, however, is the timely introduction of well-chosen geometrical probability problems into the existing sequence of topics. This procedure promotes an easy transition into probability and does not disrupt the present course structure. This article presents some interesting probability problems that may be used in this way.

The Problems

This collection of problems has been divided into four categories according to the type of geometrical figure involved: segments and lines, triangles,

175

quadrilaterals, and circles. The problems have also been classified by the level of difficulty. Those problems with two stars are very difficult, those with one star are moderately difficult, and those with no stars are easy. If a prerequisite mathematical knowledge or skill is not obvious from the content of the problem, a special note is included immediately after the problem.

It is suggested that the problems be used in three ways. For a two-star problem, the teacher can lead a group discussion concerning the problem and its complete solution. For a one-star problem, it is recommended that the teacher introduce the problem, give a few hints about a solution, and assign the completion of the problem as homework. The simple problems can be assigned as homework with no hints.

Most of the problems involve only simple notions of probability. However, prior to the introduction of the first geometrical probability problem it might be wise for the teacher to spend a few minutes reviewing fundamental probability concepts, with particular emphasis on problems in which the probability is 0 or 1.

Answers and selected hints are provided at the end along with three complete solutions that illustrate three different techniques: coordinate graphing (Segments and Lines, #4), tables (Triangles, #2), and using pictures (Triangles, #5). These techniques may also be used in many of the other problems.

Segments and Lines

1. If $AB = 2AC = 3AE = 4DB$ and if F is a point selected randomly from $\overline{AB}$, what is the probability that F is between
   a) A and C?   b) C and D?   c) E and C?

2. Points $A, B, C, D, E,$ and F are pictured in the grid below. If X and Y are distinct points selected randomly from $\{C, D, E, F\}$, what is the probability that
   a) $XY = AB$?   b) $XY > AB$?   c) $\overleftrightarrow{XY} \parallel \overleftrightarrow{AB}$?   d) $\overleftrightarrow{XY} \perp \overleftrightarrow{AB}$?

3. The lines ℓ_1 and ℓ_2 are represented by the equations $y = m_1x + 11$ and $y = m_2x + 7$, respectively, with m_1 selected randomly from
$$\left\{-\frac{4}{3}, \frac{3}{4}, \frac{1}{4}\right\}$$

and m_2 selected randomly from

$$\left\{ -\frac{3}{4}, \ -1\frac{1}{3}, \ -\frac{1}{3} \right\}.$$

What is the probability that

a) $\ell_1 \parallel \ell_2$?   b) $\ell_1 \perp \ell_2$?   c) $\ell_1 = \ell_2$?

**4. If E is the midpoint of $\overline{AB}$, C is selected randomly from $\overline{AE}$, and D is selected randomly from $\overline{EB}$, what is the probability that $CD < \frac{1}{4} AB$? (Algebraic inequality skills are required.)

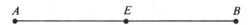

Triangles

1. Let $A = \{3, 5\}$, $B = \{4, 5, 12\}$, and $C = \{5, 13\}$. Suppose x is selected randomly from A, y from B, and z from C. What is the probability that

a) a triangle can be formed with sides of length x, y, and z?

b) an isosceles triangle (nonequilateral) can be formed with sides of length x, y, and z?

c) an equilateral triangle can be formed with sides of length x, y, and z?

d) a right triangle can be formed with sides of length x, y, and z?

2. A bag contains sticks of length 2, 3, 4, 5, and 6. If three sticks are selected randomly from the bag, what is the probability that

a) a triangle can be formed with these three sticks?

b) a right triangle can be formed with these three sticks?

c) a right triangle can be formed with these three sticks, given that the three sticks can be used to form a triangle?

3. Given $\triangle ABC$, $AB > BC$, $m\angle A = m\angle D$, and $AB = DE$. If F is selected on $\overrightarrow{DH}$ so that $BC = EF$, what is the probability that $\triangle ABC \cong \triangle DEF$? (Knowledge about the construction of a triangle, given the length of two sides and the measure of an angle opposite one of these sides, is required.)

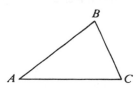 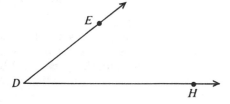

*4. Given that $\triangle ABC$ is a right triangle with hypotenuse $\overline{AB}$, $\overline{CD}$ is an altitude of $\triangle ABC$ and $AB = 25$. If AD is an integer, what is the probability that CD is an integer?

**5. If a stick is broken at random into three pieces, what is the probability that the three pieces can be used to form a triangle?

Quadrilaterals

1. In a square grid composed of three rows of three points, if two distinct points are chosen at random from the bottom row and two distinct points are chosen at random from the top row, what is the probability that the quadrilateral determined by these four points is a

   a) parallelogram?

   b) trapezoid?

   c) rectangle?

   d) square?

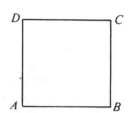

*2. Quadrilateral *ABCD* is a square with one unit on each side. Let *P* be a point chosen at random on $\overline{BC}$. Find the probability that the area of trapezoid *APCD* is greater than $\frac{2}{3}$. (Algebraic inequality skills are required.)

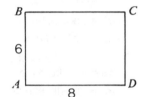

3. Quadrilateral *ABCD* is a rectangle with *AB* = 6 and *AD* = 8. If *E* is selected randomly in the interior of this rectangle, what is the probability that the area of △*AED* is—

   a) greater than 16?

   b) between 4 and 12?

   (Algebraic inequality skills are required.)

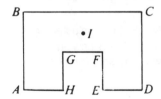

*4. In the figure, *A*, *B*, *C*, and *D* are the vertices of a rectangle; *EFGH* is a square; *AH* = *GF* = *ED* = 1; *CD* = 2; *I* is equidistant from $\overline{AB}$ and $\overline{CD}$; and *I* is equidistant from $\overline{BC}$ and $\overline{GF}$. If *J* is a point picked at random inside polygon *ABCDEFGH*, what is the probability that $\overline{IJ}$ is not entirely inside the polygon?

**5. A dartboard in the form of a square is pictured. *E* is the midpoint of $\overline{AD}$. If a dart is thrown randomly into the dartboard, what is the probability that it will be inside quadrilateral *EFCD*? (Similarity-of-triangles concepts are required.)

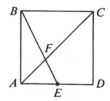

Circles

1. Three distinct points are selected at random on a circle. What is the probability that these three points are collinear?

2. In the figure, $m\overarc{AB} = 36$. Let C be any point chosen at random on the circle with $C \neq A$ and $C \neq B$. Find the probability that $m\angle BCA = 18$.

3. If C is a point selected randomly on a circle with diameter $\overline{AB}$ and such that $C \neq A$ and $C \neq B$, what is the probability that

    a) $m\angle CAB \geq 45$?

    b) $m\angle CAB \leq 30$?

    c) $\triangle ABC$ is a right triangle?

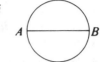

4. A target is composed of a 15-inch square and three concentric circles with radii of 3 inches, 5 inches, and 7 inches as shown.

    a) Find the probability of hitting a bull's-eye, given that the target is hit.

    b) Find the probability of hitting the target but not scoring (i.e., hitting outside the largest circle).

5. Six points on a circle are the vertices of a regular hexagon. If three points are selected at random from these six points, what is the probability that these three points will be vertices of

    a) an equilateral triangle?

    b) a right triangle?

*6. If $AB = r$, where r is the radius of a circle, and C is selected randomly on the circle, then what is the probability that $\triangle ABC$ is acute?

*7. If A and B are points selected at random on a circle of radius r, what is the probability that $AB > r$?

*8. In the figure, $\overline{AC}$ is a diameter of the circle and $AB = BC$. Suppose D is a point chosen randomly inside the circle.

    a) Find the probability that $ABCD$ (where the points are taken in order) is a quadrilateral.

    b) Find the probability that $ABCD$ is a convex quadrilateral.

    c) Find the probability that $ABCD$ is a nonconvex quadrilateral.

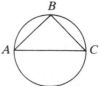

Answers

Here is a complete list of answers. Hints are included for some of the difficult problems.

Segments and Lines

1: ½, ¼, ⅙

2: ⅙, ⅔, ⅙, 0 (*Hint:* $\overline{XY} \in \{\overline{CD}, \overline{CE}, \overline{CF}, \overline{DE}, \overline{DF}, \overline{EF}\}$.)

3: ⅑, ⅑, 0

4: ⅛ (See complete solution below.)

Triangles

1: ½, ¼, 1/12, ⅙ (*Hint:* List the twelve possibilities.)

2: 7/10, 1/10, 1/7 (See complete solution below.)

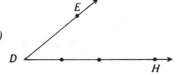

3: ½ (*Hint:* There are two possible positions for *F*.)

4: ⅙ (*Hint:* The altitude to the hypotenuse of a right triangle is the geometric mean of the segments of the hypotenuse; that is, $CD = \sqrt{AD \cdot DB}$.)

5: ¼ (See complete solution below.)

Quadrilaterals

1: 5/9, 4/9, ⅓, ⅑ (*Hint:* There are three possible line segments for each row. List the nine possibilities.)

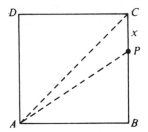

2: ⅔ (*Hint:* The probability that area $APCD > ⅔$ is equivalent to $½ + (½) (1) x > ⅔$, or $x > ⅓$. This happens ⅔ of the time.)

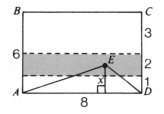

3: ⅓, ⅓

(*Hint:* For part *b*, $4 < $ area $\triangle AED < 12 \Leftrightarrow 1 < x < 3$. A similar approach will work for part *a*.)

4: $\frac{1}{5}$ (*Hint:* $\overline{IJ}$ is not entirely in the polygon whenever J is in $\triangle AHG$ or $\triangle DEF$. The area of these two triangles is $\frac{1}{5}$ of the polygon.)

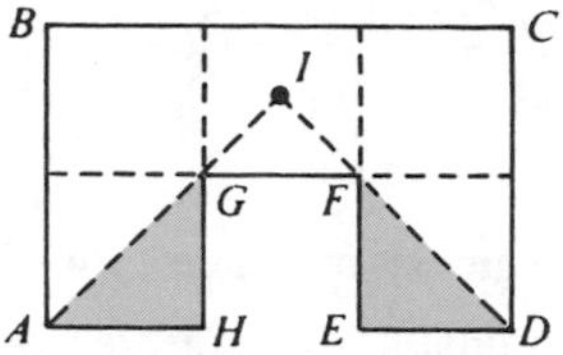

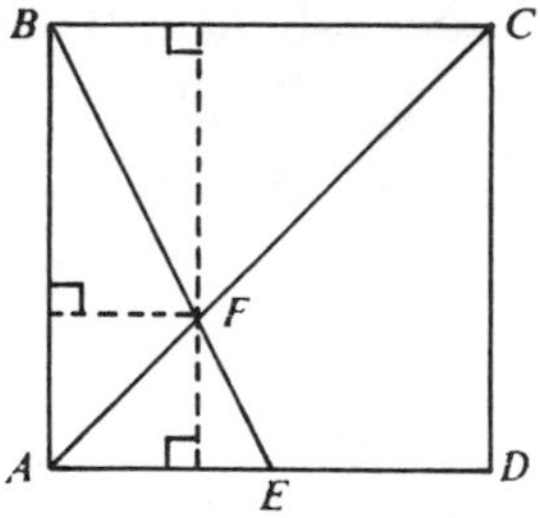

5: $\frac{5}{12}$ (*Hint:* $\triangle BCF$ is similar to $\triangle EFA$, area $\triangle ABF = 2 \cdot$ (area $\triangle AEF$), area $\triangle BCF = 4 \cdot$ (area $\triangle AEF$), area $EFCD = 5 \cdot$ (area $\triangle AEF$), and area $ABCD = 12 \cdot$ (area $\triangle AEF$).)

Circles

1: 0

2: $\frac{9}{10}$

3: $\frac{1}{2}$, $\frac{1}{3}$, 1

4: $\pi/25$, $(225 - 49\pi)/225$

5: $\frac{1}{10}$, $\frac{2}{5}$ (*Hint:* List all possible triangles.)

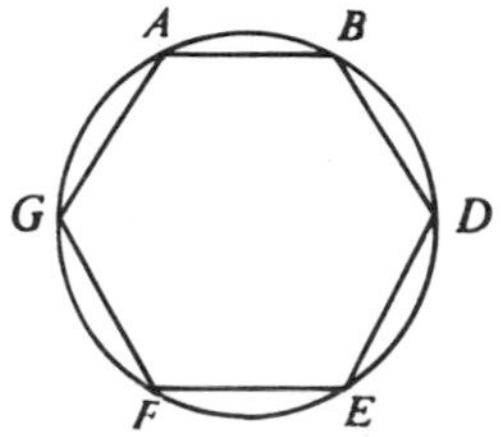

6: $\frac{1}{6}$ (*Hint:* Inscribe regular hexagon $ABDEFG$ in the circle. $\triangle ABC$ is acute whenever C is on $\widehat{FE}$; that is, $\frac{1}{6}$ of the time.)

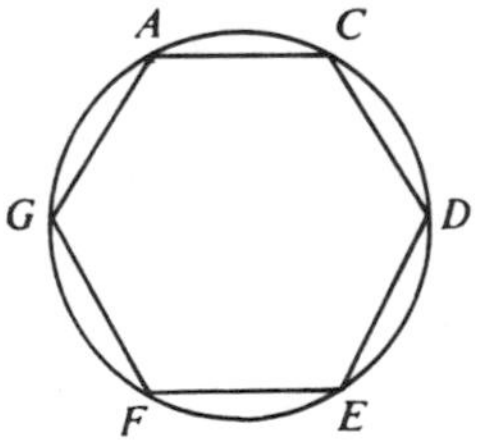

7: $\frac{2}{3}$ (*Hint:* Let A be any fixed point on the circle and inscribe regular hexagon $ACDEFG$. $AB > r$ whenever B is on major arc CG and $B \neq C, G$. Thus, $\frac{4}{6}$ of the time.)

8: $(1+\pi)/(2\pi)$, $\frac{1}{2}$, $1/\pi$ (*Hint:* For part *a,* D cannot be in sectors cut off by $\overline{AB}$ or $\overline{BC}$; for part *b,* D must be in the "lower" semicircle; and for part *c,* D must be in $\triangle ABC$.

Solutions

Segments and Lines, #4

Establish a coordinate system on $\overleftrightarrow{AB}$ so that the coordinate for A is 0 and the coordinate for B is 1. Let the coordinate for C be x and the coordinate for D be y.

The conditions of the problem require that

$$0 < x < \tfrac{1}{2} \text{ and } \tfrac{1}{2} < y < 1.$$

The graph of the set of all ordered pairs (x, y) that meet these conditions is the interior of square $FGHI$ pictured below.

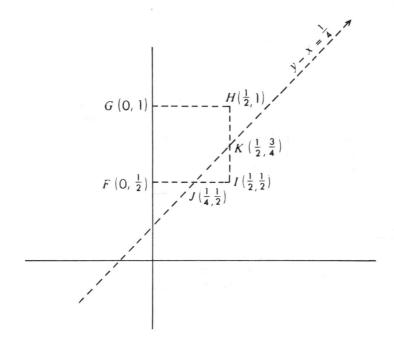

Since $CD = y - x$ and $AB = 1$, the condition $CD < (\tfrac{1}{4}) AB$ is equivalent to

$$y - x < \tfrac{1}{4}.$$

The condition $y - x < \tfrac{1}{4}$ requires that (x, y) be in the interior of $\triangle JIK$. Since the area of $\triangle JIK$ is $\tfrac{1}{8}$ of the area of square $FGHI$, the required probability is $\tfrac{1}{8}$.

Triangles, #2

The table below provides a systematic way of showing the possible ways that the sticks can be selected.

3 sticks	Triangle?	Right triangle?
2, 3, 4	Yes	No
2, 3, 5	No	No
2, 3, 6	No	No
2, 4, 5	Yes	No
2, 4, 6	No	No
2, 5, 6	Yes	No
3, 4, 5	Yes	Yes
3, 4, 6	Yes	No
3, 5, 6	Yes	No
4, 5, 6	Yes	No

Triangles, #5

Let $\overline{AB}$ represent the stick, let D be the left break point, and let E represent the right break point. Furthermore, let C be a point such that $\triangle ABC$ is an equilateral triangle, let G, H, and I be the midpoints of $\overline{AB}$, $\overline{BC}$, and $\overline{CA}$, and let F be the intersection of a line through D parallel to $\overline{AC}$ and a line through E parallel to $\overline{BC}$. Then F is a uniquely determined point in the interior of $\triangle ABC$.

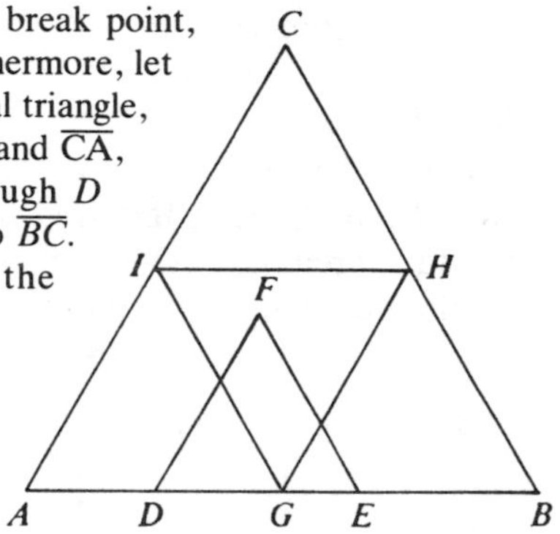

If F is in the interior of $\triangle GHI$, then a triangle can be formed (triangle inequality), but if F is in the interior of $\triangle AGI$, $\triangle GHB$, or $\triangle HIC$, the triangle inequality does not hold and no triangle can be formed. Hence, the required probability is ¼.

BIBLIOGRAPHY

Dahlke, Richard, and Robert Fakler. "Geometrical Probability." In *Teaching Statistics and Probability*, 1981 Yearbook of the National Council of Teachers of Mathematics, pp. 143–53. Reston, Va.: The Council, 1981.

Woodward, Ernest, and Jim Ridenhour. "An Interesting Probability Problem." *Mathematics Teacher* 75 (December 1982): 765–68.

Mathematical Applications of Geometry

Vincent P. Schielack, Jr.

Eʟᴇᴍᴇɴᴛᴀʀʏ properties of geometrical objects such as rectangles, triangles, and circles can be used to justify many formulas in other branches of
mathematics, at least on an intuitive basis. These justifications rely on
familiar concepts to convince the student of the validity of the formulas;
moreover, the justifications show quite clearly the pervasiveness of geometry
in mathematics. Often these arguments consist simply of a self-explanatory
diagram. The intent of this expository article is to demonstrate some of
these applications of geometry to algebra, combinatorics, analysis, and analytic geometry. References for these justifications are included. I have not
included proofs of trigonometric identities, since I consider trigonometry a
subfield of geometry.

Algebra

The distributive law of multiplication over addition. This law is demonstrated by comparing the area of the large rectangle in figure 16.1 to the
sum of the areas of the smaller rectangles. My students never fail to appreciate this geometric light on the distributive law.

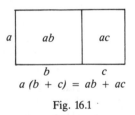

$$a (b + c) = ab + ac$$

Fig. 16.1

Multiplication of signed numbers (Lariviere 1957). The four graphs in
figure 16.2 represent the four cases for the multiplication of signed numbers.
Applications of properties of parallel lines and similar triangles yield the
signs of the results. In each, we have a pair of coordinate axes oriented in
the usual way. *OU* represents unit length in the positive *x* direction. *OA* on
the *x*-axis represents (in magnitude and sign) the first quantity to be multiplied. *OB* on the *y*-axis represents (in magnitude and sign) the second

quantity to be multiplied. $\overline{BU}$ is then drawn, followed by the line parallel to $\overline{BU}$ through A, which intersects the y-axis at point C. Since $\triangle OAC$ is similar to $\triangle OUB$ in each graph, we have $\dfrac{OC}{OB} = \dfrac{OA}{OU}$, or $OC = \dfrac{OA \cdot OB}{OU} = OA \cdot OB$, since $OU = +1$. Thus, OC represents the product of the two signed numbers OA and OB, and the sign of OC determines the sign of this product.

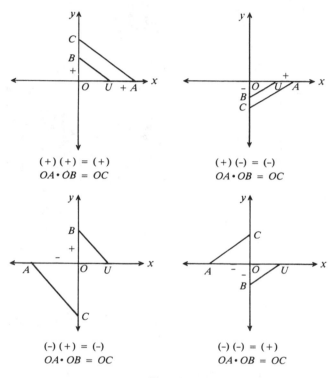

Fig. 16.2

Binomial products (Horak and Horak 1981). The binomial products $(a + b)(c + d)$, $(a + b)^2 (a - b)^2$, and $(a + b)(a - b)$ can all be justified intuitively by dissecting rectangular areas. Figure 16.3 illustrates the first three products. (Note that $(a + b)^2$ is a special case of the first.) The identity $(a + b)(a - b) = a^2 - b^2$ can be established in two distinct ways, as shown in figure 16.4, depending on which side of the equation one begins with. Note that figure 16.4a actually represents a factoring formula, whereas figure 16.4b represents a product formula.

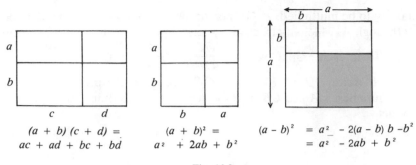

Fig. 16.3

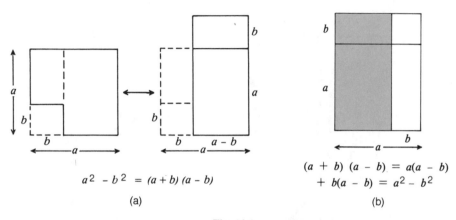

$a^2 - b^2 = (a + b)(a - b)$

(a)

$(a + b)(a - b) = a(a - b)$
$+ b(a - b) = a^2 - b^2$

(b)

Fig. 16.4

Completing the square (Gallant 1983; Moore 1978). Figure 16.5 shows geometrically the process used in completing the square, yielding the familiar result $x^2 + ax = \left(x + \dfrac{a}{2}\right)^2 - \left(\dfrac{a}{2}\right)^2$. Note that the illustration lends a completely new meaning to the term *completing the square*.

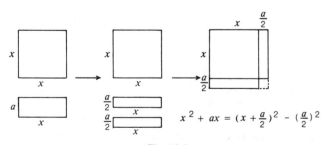

$x^2 + ax = (x + \frac{a}{2})^2 - (\frac{a}{2})^2$

Fig. 16.5

Combinatorics

Students studying mathematical induction are often asked to prove the closed formulas for the sums of the first n positive integers, squares of positive integers, odd positive integers, and cubes of positive integers. Although the results are easily proved by induction, students naturally wonder how the closed expressions are obtained. The following demonstrations show the patterns that indicate what the closed formulas should be.

Sum of the first n positive integers. In figure 16.6, two congruent arrangements of unit squares are used to obtain the familiar result $1 + 2 + \ldots + n = \dfrac{n(n + 1)}{2}$ by area dissection. (For a modification of this proof, see Richards [1984].)

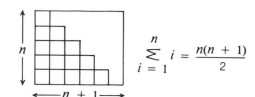

$$\sum_{i = 1}^{n} i = \frac{n(n + 1)}{2}$$

Fig. 16.6 ←——$n + 1$——→

Sum of the squares of the first n positive integers (Siu 1984). Here the integers are represented as unit cubes. (See fig. 16.7.) A clever arrangement of three congruent sets of these unit cubes yields a rectangular parallelepiped, and the result $1^2 + 2^2 + \ldots + n^2 = \dfrac{1}{3} n (n + 1) (n + \dfrac{1}{2}) = \dfrac{n(n + 1)(2n - 1)}{6}$ is obtained by volume dissection.

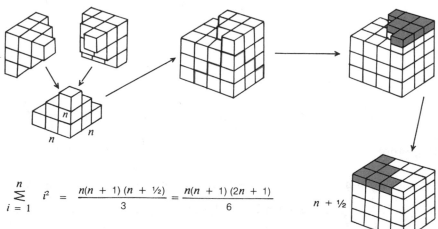

$$\sum_{i = 1}^{n} i^2 = \frac{n(n + 1)(n + \frac{1}{2})}{3} = \frac{n(n + 1)(2n + 1)}{6}$$

Fig. 16.7

Sum of the first n odd positive integers (Salkind 1958). We present two proofs of the result $1 + 3 + 5 + \ldots + (2n - 1) = n^2$. The first of these, as shown in figure 16.8, is an area dissection proof similar to that for the sum of the first n positive integers. (In fact, a general proof can be devised similarly for the sum of the first n terms of any arithmetic sequence.) The second (fig. 16.9) is also an area dissection proof, but it involves circular instead of square regions.

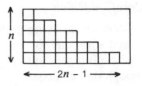

$$\sum_{i=1}^{n} (2i - 1) = \frac{n(2n)}{2} = n^2$$

Fig. 16.8

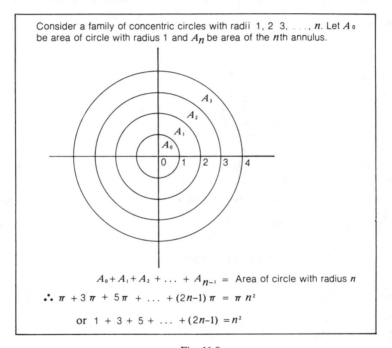

Consider a family of concentric circles with radii 1, 2 3, . . . , n. Let A_0 be area of circle with radius 1 and A_n be area of the nth annulus.

$A_0 + A_1 + A_2 + \ldots + A_{n-1}$ = Area of circle with radius n

$\therefore \pi + 3\pi + 5\pi + \ldots + (2n-1)\pi = \pi n^2$

or $1 + 3 + 5 + \ldots + (2n-1) = n^2$

Fig. 16.9

Sum of the cubes of the first n positive integers (Salkind 1958; Love 1977). One proof of the formula $1^3 + 2^3 + \ldots + n^3 = \left(\frac{n(n + 1)}{2}\right)^2$ is based on the same idea used in the last proof. Instead of using concentric circles of radii 1, 2, 3, . . ., n, we use circles with radii 1, 3, 6, . . ., $\frac{n(n + 1)}{2}$. (See fig. 16.10.)

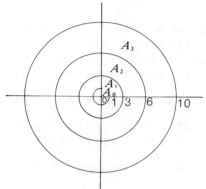

$$A_0 + A_1 + A_2 + A_3 + \ldots + A_{n-1} = \text{Area of circle with radius } \frac{n(n + 1)}{2}$$

$$\therefore 1^3 \pi + 2^3 \pi + 3^3 \pi + \ldots + n^3 \pi = \pi \left(\frac{n(n + 1)}{2} \right)^2$$

$$\text{or} \quad 1^3 + 2^3 + 3^3 + \ldots + n^3 = \left(\frac{n(n + 1)}{2} \right)^2$$

Fig. 16.10

An alternative proof represents the cube of the positive integer n as n times the area of an $n \times n$ square. In figure 16.11, start at the upper left corner and notice that the area representing 1^3 can be joined together with the area representing 2^3 to form a square of side $(1 + 2)$. This square can be joined with the area representing 3^3 to form a square of side $(1 + 2 + 3)$. Continuing in this fashion and comparing resultant areas, we obtain $1^3 + 2^3 + \ldots + n^3 = (1 + 2 + \ldots + n)^2$. (See Fry [1985] for a related proof involving volume.)

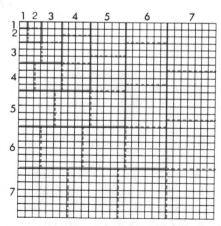

Fig. 16.11 $1^3 + 2^3 + \ldots + n^3 = (1 + 2 + \ldots + n)^2$

Analysis

The arithmetic-geometric mean inequality (Gallant 1977; Eddy 1985). (Note how many geometric results are used in this proof.) The arithmetic-geometric mean inequality states that if a and b are positive, the geometric mean of a and b is no greater than their arithmetic mean; that is, $\sqrt{ab} \leq \frac{a + b}{2}$. One proof is embodied in figure 16.12. Line segments $\overline{PN}$ and $\overline{NM}$ of lengths a and b, respectively, are joined to form $\overline{PM}$, of length $a + b$. The midpoint O of this segment can be used to construct the semicircle shown, which has radius $\frac{a + b}{2}$. In particular, $OR = \frac{a + b}{2}$. Triangle PQM is a right triangle, and thus $QN = \sqrt{ab}$, the geometric mean of a and b. Since $QN \leq OR$, we have $\sqrt{ab} \leq \frac{a + b}{2}$.

Another extremely elegant proof of this result depends solely on the Pythagorean theorem. It is shown in figure 16.13.

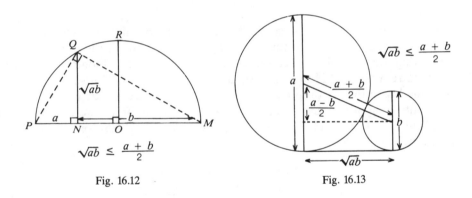

Fig. 16.12   Fig. 16.13

An inequality involving four means (Ercolano 1972). We can extend the arithmetic-geometric mean inequality to include the harmonic mean and the root-mean square as well. The harmonic mean of positive numbers a and b is $\frac{2ab}{a + b}$, and the root-mean square is $\sqrt{\frac{1}{2}(a^2 + b^2)}$. Let us denote the harmonic mean, geometric mean, arithmetic mean, and root-mean square of a and b by H.M., G.M., A.M., and R.M., respectively. Consider figure 16.14. We showed previously that G.M. $= QN$ and A.M. $= OR$. We will show that H.M. $= SQ$ and R.M. $= NR$. Triangle SQN is similar to $\triangle NQO$, so that

$$\frac{SQ}{QN} = \frac{QN}{QO}, \text{ or } SQ = \frac{(QN)^2}{QO} = \frac{(\sqrt{ab})^2}{\dfrac{a+b}{2}} = \frac{2ab}{a+b}.$$

$$NR = \sqrt{(OR)^2 + (NO)^2} = \sqrt{\left(\frac{a+b}{2}\right)^2 + \left(\frac{a-b}{2}\right)^2} = \sqrt{\frac{1}{2}(a^2 + b^2)}.$$

From figure 16.15 it is obvious that $SQ \leq QN \leq OR \leq NR$, or H.M. $\leq$ G.M. $\leq$ A.M. $\leq$ R.M. (Ercolano [1972] also contains another diagram illustrating this inequality.)

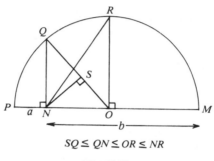

$$SQ \leq QN \leq OR \leq NR$$

Fig. 16.14

Analytic Geometry

The distance from a point to a line (Eisenman 1969). (Anyone who has derived this formula using the standard analytic geometry methods should be very impressed with this proof using similar triangles.) Consider the distance from point P with coordinates (a,c) to the line $y = mx + b$. (See fig. 16.15.) If $\overline{OP}$ is a vertical segment, then O has coordinates $(a, ma + b)$, and $OP = |ma + b - c|$. Triangle QRS is formed to display the slope of $y = mx + b$; that is, the horizontal segment $\overline{QR}$ has length 1, and the vertical segment $\overline{SR}$ has length $|m|$. Noting that $\triangle NPO$ is similar to $\triangle RQS$, we have $d = \dfrac{d}{1} = \dfrac{|ma + b - c|}{\sqrt{1 + m^2}}$. Using the standard form $Ax + By + C$ of the equation of the line rather than the slope-intercept form transforms the formula into the more familiar

$$d = \frac{|Aa + Bc - C|}{\sqrt{A^2 + B^2}}.$$

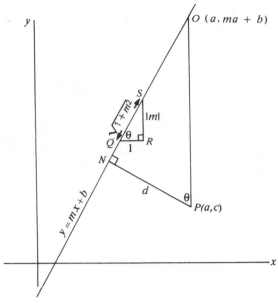

Fig. 16.15

Slopes of perpendicular lines (Munem and Foulis 1982). A necessary and sufficient condition for two lines to be perpendicular is that their slopes be negative reciprocals of each other. A proof by standard analytic geometry techniques can be quite tedious; here we use instead the Pythagorean theorem and its converse. Let L_1 and L_2 be two perpendicular lines with slopes m_1 and m_2, respectively. Translating the origin O to the point of intersection of L_1 and L_2 does not affect the angle between the lines or their slopes. (See fig. 16.16.) Let P be the point $(1, m_1)$ on L_1 and let Q be the point $(1, m_2)$ on L_2. By the Pythagorean theorem and its converse, $\triangle POQ$ is a right triangle (and L_1 and L_2 are perpendicular) if and only if $(OP)^2 + (OQ)^2 = (PQ)^2$. Using the distance formula, we find $(OP)^2 = 1 + m_1^2$, $(OQ)^2 = 1 + m_2^2$, and $(PQ)^2 = m_1^2 - 2m_1m_2 + m_2^2$. Thus, $(OP)^2 + (OQ)^2 = (PQ)^2$ if and only if $(1 + m_1^2) + (1 + m_2^2) = m_1^2 - 2m_1m_2 + m_2^2$, which reduces to $m_1m_2 = -1$. (Wahlgren [1977] and Dunkels [1977] present proofs of this property through the consideration of similar triangles.)

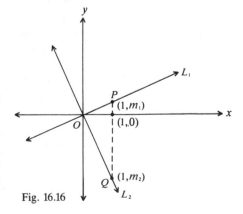

Fig. 16.16

These are just a few examples of the diverse and powerful applications of elementary geometry to other branches of mathematics. It is hoped that classroom teachers will take note of some of these applications and pass them along to their students. An intuitive geometric demonstration can literally be worth a thousand words to the mathematics student.

REFERENCES

Dunkels, Andrejs. "Another Slant on Slope." *Mathematics Teacher* 70 (November 1977): 644–45.

Eddy, Roland H. "Behold! The Arithmetic-Geometric Mean Inequality." *College Mathematics Journal* 16 (June 1985): 208.

Eisenman, R. L. "An Easy Way from a Point to a Line." *Mathematics Magazine* 42 (January 1969): 40–41.

Ercolano, Joseph L. "Geometric Interpretations of Some Classical Inequalities." *Mathematics Magazine* 45 (September 1972): 226.

Fry, Alan L. "Proof without Words: Sum of Cubes." *Mathematics Magazine* 58 (January 1985): 11.

Gallant, Charles. "Proof without Words: A Truly Geometric Inequality." *Mathematics Magazine* 50 (March 1977): 98.

————. "Proof without Words: Completing the Square." *Mathematics Magazine* 56 (March 1983): 110.

Horak, Virginia M., and Willis J. Horak. "Geometric Proofs of Algebraic Identities." *Mathematics Teacher* 74 (March 1981): 212–16.

Lariviere, R. "Geometric Reinforcement." *Mathematics Magazine* 30 (March/April 1957): 205–6.

Love, J. Barry. "Proof without Words: Cubes and Squares." *Mathematics Magazine* 50 (March 1977): 74.

Moore, Charles G. "Completing the Square—a Laboratory Approach." *Two-Year College Mathematics Journal* 9 (September 1978): 215–18.

Munem, Mustafa M., and David J. Foulis. *College Algebra with Applications.* New York: Worth, 1982.

Page, Warren. "Proof without Words: Geometric Sums." *Mathematics Magazine* 54 (September 1981): 201.

Richards, Ian. "Proof without Words: Sum of Integers." *Mathematics Magazine* 57 (March 1984): 104.

Salkind, Charles T. "A Geometric Viewpoint in Elementary Series." *Mathematics Teacher* 51 (May 1958): 343.

Siu, Man-Keung. "Proof without Words: Sum of Squares." *Mathematics Magazine* 57 (March 1984): 92.

Van Beynen, John, and Robert L. McGinty. "A Geometric Interpretation of Series." *Mathematics Teacher* 74 (March 1981): 218–21.

Wahlgren, Jack C. "A Note on Perpendicular Lines and Slopes." *Mathematics Teacher* 70 (January 1977): 64.

ADDITIONAL READING

(The following works, though not cited in the article, are additional sources of applications of geometric models to other branches of mathematics.)

Buck, Frank. "Behold! The Midpoint Rule Is Better Than the Trapezoid Rule for Concave Functions." *College Mathematics Journal* 16 (January 1985): 56.

Chilaka, James O. "Proofs without Words: Combinatorial Identities." *Mathematics Magazine* 52 (September 1979): 206.

—————. "Proof without Words: Sum of Squares." *Mathematics Magazine* 56 (March 1983): 90.

De Noya, Louis E. "The Geometric Progression Presented Geometrically." *Mathematics Teacher* 56 (March 1963): 146–47.

Garfunkel, J., and B. Plotkin. "Using Geometry to Prove Algebraic Inequalities." *Mathematics Teacher* 59 (December 1966): 730–34.

Golomb, Solomon W. "Proof without Words: A 2 × 2 Determinant Is the Area of a Parallelogram." *Mathematics Magazine* 58 (March 1985): 107.

Hoehn, Larry. "A Geometrical Interpretation of the Weighted Mean." *College Mathematics Journal* 15 (March 1984): 135–39.

Kazarinoff, Nicholas D. *Geometric Inequalities*. New York: Random House, 1961.

Killam, S. Douglas. "A Few Graphical Methods." *Mathematics Teacher* 6 (September 1913): 10–16.

Landauer, Edwin G. "Proof without Words: Square of an Even Positive Integer." *Mathematics Magazine* 58 (September 1985): 236.

—————. "Proof without Words: Square of an Odd Positive Integer." *Mathematics Magazine* 58 (September 1985): 203.

Meserve, B. E. "Using Geometry in Teaching Algebra." *Mathematics Teacher* 45 (December 1952): 567–71.

Monzingo, M. G. "A Visual Interpretation of Independent Events." *Two-Year College Mathematics Journal* 13 (June 1982): 197–98.

Nannini, Amos. "Geometric Solution of a Quadratic Equation." *Mathematics Teacher* 59 (November 1966): 647–49.

Page, Warren. "Proofs without Words: Count the Dots." *Mathematics Magazine* 55 (March 1982): 97.

Schaefer, Sister M. Geralda. "Geometric Representation of Some Integer Sums." *Texas Mathematics Teacher* 32 (January 1985): 14–15.

Schattschneider, Doris. "Proof without Words: The Arithmetic Mean–Geometric Mean Inequality." *Mathematics Magazine* 59 (February 1986): 11.

Turner, Barbara. "A Geometric Proof That $\sqrt{2}$ Is Irrational." *Mathematics Magazine* 50 (November 1977): 263.

Wakin, Shirley. "Proof without Words: Algebraic Areas." *Mathematics Magazine* 57 (September 1984): 231.

17

Geometry for Calculus Readiness

Richard H. Balomenos
Joan Ferrini-Mundy
Thomas Dick

Geometry! Why do I need to study it?" How many times have teachers of geometry heard that question as a class is about to embark on a yearlong course in geometry? All too often the answers are dutifully given about the importance of geometry as an example of a deductive system, one of the marvels of mathematics inherited from ancient Greece. We stress the need to understand logical reasoning and its relation to axioms, theorems, and proofs, as illustrated by Euclidean geometry. In addition, many teachers view high school geometry as an isolated course having little bearing on later courses, especially college mathematics—algebra and trigonometry are seen as the courses that are central to future success in mathematics. In fact, some teachers see little relationship between the geometry experience and the development of problem-solving facility, and they let their students know it!

There is more to the answer to the question, "Why do I need to study geometry?" Evidence is growing that many of our students' difficulties in calculus are due to poor preparation in geometry. We suggest an expanded role for geometry in the high school: the study of geometry should also provide readiness for calculus and develop spatial visualization.

Geometry, Spatial Ability, and Calculus Readiness

Certain concepts of the traditional college calculus course are frequently introduced through geometric representations, and a majority of the physical problems in calculus are virtually impossible to understand without adequate visual representations. Many of the ideas needed for understanding college calculus are based on the traditional secondary school geometry course, but college calculus students are sometimes taken by surprise when somewhere in the middle of a complicated calculus problem or the explanation of a concept reference is made to similar triangles, the Pythagorean theorem, or the volume of a cylinder. It would help students if we could better prepare them for these occurrences of geometric ideas in new contexts, and this is possible in the secondary school geometry course.

In order to solve many of the applied problems traditionally included in calculus, the ability to construct a pictorial representation of some geometric configuration, based on a complicated verbal description, is essential. The typical secondary school geometry course does not impose this type of spatial demand, and as a result students greet the rather complicated figures and diagrams that they see in calculus with some consternation.

A number of the fundamental concepts in calculus, such as the definite integral, the derivative, the area between curves, the volume of solids of revolution, and the surface area of solids, frequently are introduced with heavy reliance on pictures. Despite the calculus teacher's predilection for diagrams, our research indicates that students resist the use of geometric and spatial strategies in actually solving calculus problems. The blind use of a strictly analytic formula can, of course, lead to disastrous results. For example, here is a problem that many students fail to solve correctly because of an incomplete analysis of the situation:

> Find the area enclosed by the graphs of $y = x^3$ and $y = x$ between $x = -1$ and $x = 1$.

Students who evaluate the definite integral

$$\int_{-1}^{1} (x^3 - x)dx \text{ or } \int_{-1}^{1} (x - x^3) \, dx$$

usually do so in the belief that the expression

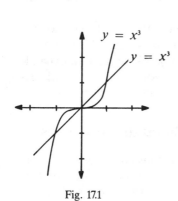

$$\int_{a}^{b} [f(x) - g(x)]dx$$

will give the area enclosed by $y = f(x)$ and $y = g(x)$ between $x = a$ and $x = b$. However, this formula is valid only if $f(x) \geq g(x)$ for $a \leq x \leq b$. The student who graphs the functions involved (see fig. 17.1) is often alerted to the need of evaluating

Fig. 17.1

$$\int_{-1}^{0} (x^3 - x)dx + \int_{0}^{1} (x - x^3)dx$$

as the expression giving the correct result. Without graphing, it is doubtful that a student would recognize that symmetry could also be exploited to give

$$2\int_{0}^{1} (x - x^3)dx$$

as an expression for this area.

Specific areas of calculus rely heavily on geometric representations. To translate ladders sliding down walls into meaningful geometric representations, or to represent sand falling onto a conical pile, students must formulate two-dimensional representations of three-dimensional dynamic situations. Following are suggestions for anticipating these areas in secondary school geometry.

Ideas for Secondary School Geometry

The secondary school geometry course should explicitly develop the capacity of students to formulate geometric representations in unfamiliar contexts and require them to use visual strategies that demand the synthesis of previously disparate geometric approaches (e.g., finding similar triangles in the cross section of an inscribed cone). We have selected four topics in calculus in which our students have particular difficulty because of the geometric and visual elements involved: maximum-minimum problems, related-rate problems, volume-of-solids-of-revolution problems, and other applied problems. In discussing each topic we attempt to highlight the ways in which these areas of calculus could be "anticipated" through specific emphases and examples that could be included in the secondary school geometry course. Calculus teachers also might find these topics well worth discussing from a geometric point of view immediately before their introduction.

A Geometric Look at Maximum-Minimum Problems

One of the more important applications of the calculus is its use in solving problems of *optimization:* under given constraints how do we maximize or minimize certain quantities? Most students appreciate this use of calculus, since they are well aware of the practical concerns of maximizing benefits for given cost or minimizing costs to realize desired benefits. However, students often view "max-min" problems as quite difficult. There are a large number of max-min problems in which the students' difficulties can be traced to inadequate geometric skills rather than inadequate calculus skills. As an example, consider the following two calculus problems:

1. Given that x lies in the closed interval $[0,3]$, find the value of x that maximizes the quantity V, where

$$V = (9 + 3x - x^2 - \frac{x^3}{3}).$$

2. Find the dimensions of the right circular cone of maximum volume that can be inscribed in a sphere of radius 3 units.

Students generally do not find problem 1 very difficult, but many have a

great deal of trouble with exercises like problem 2. The calculus used to
solve problem 2 is no more difficult than that re-
quired for problem 1. Indeed, if we consider the
diagram in figure 17.2 illustrating the situation described in problem 2, we see that the two problems
are actually equivalent with respect to calculus. The
radius r of the base of the cone satisfies the Pythagorean relationship $r^2 = 3^2 - x^2$, and the height
h of the cone is $h = 3 + x$. Since the volume V of
the cone is $V = \frac{1}{3}\pi r^2 h$, we next try to maximize V
as a function of x.

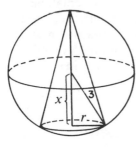

Fig. 17.2

$$V = \frac{1}{3}\pi(9 - x^2)(3 + x) \quad = \pi(9 + 3x - x^2 - \frac{x^3}{3})$$

This is the same expression that appears in problem 1. Furthermore, since
$0 \le x \le 3$ (cones with a height less than 3 units would certainly not give us
maximum volume), the problems can be considered entirely equivalent.

Students have very little trouble applying calculus to finding the maximum
of a function of one variable. Their real difficulties lie in finding the function!
The crucial steps in the solution of a max-min problem are those involving
the use of the constraints to rewrite the quantity to be optimized in terms
of one variable. Because many of the standard max-min problems in calculus are based on geometric representations, students could benefit greatly
from practice on precalculus problems that focus on the geometric strategies
typically needed in the max-min exercises.

Calculus problem: A sphere of radius 4 is inscribed in a right circular cone.
Find the dimensions of the cone with minimum volume.

Here the crucial step is to use the fact that the inscribed sphere is of radius
4 to express the volume of the cone as a function of one variable. To focus
attention on this phase of the problem, it could be rewritten for geometry
students as follows.

Geometry problem: A sphere of radius 4 is inscribed in a right circular cone.
Express the volume of the cone as a function of its height, h.

We suggest that students be encouraged to solve problems such as this by
attending carefully to their geometric components. The
basic situation is illustrated in figure 17.3. Students can
profit from drawing two-dimensional representations of
the three-dimensional situations that arise in calculus.
Geometrically, a key step in solving this problem is developing an image of the cross section. Next, students
need to "disembed" the cross section and search for the

Fig. 17.3

relevant attributes. Recognizing that the radius $\overline{DE}$ is a useful auxiliary line segment requires both the anticipation of the use of similar triangles and the visualization of the cross section within the solid (fig. 17.4).

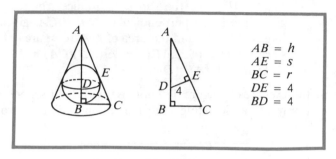

$$AB = h$$
$$AE = s$$
$$BC = r$$
$$DE = 4$$
$$BD = 4$$

Fig. 17.4

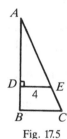

Fig. 17.5

A common student error is to visualize incorrectly and arrive at the drawing in figure 17.5. Unfortunately, $\overline{DE}$ in this drawing is not a radius of the sphere. When the correct representation in figure 17.4 is used, $\triangle ABC \sim \triangle AED$, and hence

$$\frac{r}{4} = \frac{h}{s} = \frac{r + s}{h - 4}.$$

From this it is an algebraic exercise to express r^2 as

$$\frac{16h}{h - 8} \quad \text{and} \quad V = \frac{1}{3}\pi \, \frac{16h^3}{h^2 - 8h}.$$

A slightly different class of max-min problems is represented by the following precalculus problem; this typically appears in the calculus setting, and students assume calculus is required (fig. 17.6).

Problem: Of all triangles having two given sides, show that the triangle of maximum area is the one in which the two given sides include a right angle.

Solution: Creating a representation for this problem that suggests a method of solution requires the student to envision the various angles formed as the given sides are separated.

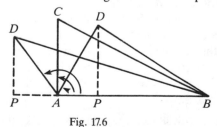

Fig. 17.6

Formulating the diagram from the verbal statement is a crucial skill in many calculus problems, as is organizing the verbal statement into what is given and what is to be proved.

Given: $AC = AD$, $AB = AB$, $\angle BAC$ is a right angle

Prove: Area of $\triangle ABC \geq$ area of $\triangle ADB$

Proof: From D we construct the perpendicular $\overline{DP}$ to line $\overrightarrow{AB}$. $DA > DP$, since the shortest distance from a point to a given line is the perpendicular. $DA = CA$, given. Therefore, $CA \geq DP$ and the area of $\triangle ABC \geq$ area of $\triangle ADB$, since the area of $\triangle ABC = \frac{1}{2}(AB)(\overline{CA})$ and the area of $\triangle ADB = \frac{1}{2}(AB)(DP)$.

We suggest the exercises in figure 17.7 for use with high school geometry students as a way of "anticipating" the maximum-minimum problems of calculus.

Geometry Problems Anticipating Max-Min Calculus Problems

1. Find the area of a rectangle inscribed in a semicircle as a function of its width.
2. Find the area of a rectangle of perimeter L as a function of its width.
3. Find the volume and surface area of a cube as functions of its diagonal length.
4. Prove: Of all parallelograms with given area and base, the rectangle has minimum perimeter.
5. Prove: Of all rectangles with equal area, the square has the minimum perimeter.
6. Prove: Of all triangles of a given base and altitude, the isosceles triangle has the minimum perimeter.
7. Prove: Of all triangles having a given base and a given perimeter, the isosceles triangle has maximum area.
8. Prove: Of all triangles that can be inscribed in a given circle, the equilateral triangle has both the greatest perimeter and the greatest area.

Fig. 17.7

A Geometric Look at Related-Rate Problems

In calculus, students frequently encounter problems in which one quantity is related to another and both change with time. Taking the derivative of each with respect to time yields a relation between their rates of change. Such problems are called related-rate problems. Setting up and solving related-rate problems is difficult for calculus students. It is the ability to develop a geometric representation of the physical situation from the complicated verbal description that is the real challenge. In many related-rate problems, the key to the solution is solving a geometry problem in which time is "frozen." Consider the following example:

A train is traveling at 60 miles an hour (or 88 feet a second) on a track 50 feet below a long, level highway bridge. A car traveling at 45 miles an hour (or 66 feet a second) passes directly over the train.

The calculus problem is "How fast are the train and car separating 10 seconds later?" The geometry problem is "How far apart are the train and car 10 seconds later? The solution in either case requires the use of the Pythagorean theorem in three dimensions. If we freeze the action at 10 seconds after the car passes directly over the train, the car will have traveled 660 feet and the train 880 feet (fig. 17.8).

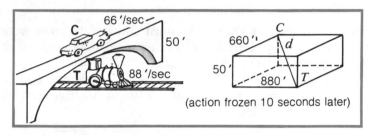

(action frozen 10 seconds later)

Fig. 17.8

$$d = \sqrt{660^2 + 880^2 + 50^2} = 10\sqrt{11^2 \cdot 6^2 + 11^2 \cdot 8^2 + 5^2}$$
$$= 10\sqrt{11^2(36 + 64) + 25} = 10\sqrt{25(11^2 \cdot 4 + 1)}$$
$$= 50\sqrt{485} \approx 1100$$

Thus the car and train will be approximately 1100 feet apart.

The spatial demands of such a problem are considerable, and when the student is also struggling with the calculus concepts involved, the situation can be overwhelming. By focusing only on the geometry of the problem, we can encourage the student's facility in finding a visual representation, and the task of finding the appropriate right triangle in this three-dimensional representation can be dealt with separately.

The following is a calculus problem in which the properties of similar triangles are used in the solution.

> Water is pouring into a conical cistern at the rate of 8 cubic feet per minute. If the height of the inverted cone is 12 feet and the radius of its circular opening is 6 feet, how fast is the water level rising when the water is 4 feet deep? (Purcell 1978, p. 160)

By "freezing the action," we convert the problem into a geometry problem: What is the volume of the water when the water is 4 feet deep? The formula for the volume of water is

$$V = \frac{1}{3}\pi r^2 h.$$

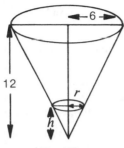

Fig. 17.9

The geometry question is to find V when $h = 4$. From similar triangles (see fig. 17.9):

$$\frac{r}{h} = \frac{6}{12}, \text{ or } r = \frac{1}{2}h \qquad V = \frac{1}{3}\pi\left(\frac{h}{2}\right)^2 h = \frac{\pi h^3}{12}$$

Therefore, when $h = 4$, $V = \dfrac{\pi 4^3}{12} = \dfrac{16\pi}{3}$ cubic feet.

Notice that here again the student must visualize the cross section of the cone, label it appropriately in terms of given relationships, and choose the appropriate similar triangles.

In figure 17.10 are some geometry exercises adapted from related-rate problems as found in any standard calculus textbook; we have used texts by Swokowski (1984) and Anton (1984).

Geometry Problems from Related-Rate Problems in Calculus

1. A ladder 20 ft long leans against a vertical building. If the bottom of the ladder slides away from the building horizontally at a rate of 3 ft/sec, how far is the bottom of the ladder from the building when the top of the ladder is 8 ft from the ground? [Swokowski]

2. At 1:00 p.m., ship A is 25 mi due south of ship B. If ship A is sailing west at a rate of 16 mi/hr and ship B is sailing south at a rate of 20 mi/hr, what is the distance between the ships at 1:30 p.m.? [Swokowski]

3. A girl starts at a point A and runs east at a rate of 10 ft/sec. One minute later, another girl starts at A and runs north at a rate of 8 ft/sec. What is the distance between them 1 minute after the second girl starts? [Swokowski]

4. The ends of a water trough 8 ft long are equilateral triangles whose sides are 2 ft long. Find the volume of the water in the trough when the depth is 8 in. [Swokowski]

5. A softball diamond has the shape of a square with sides 60 ft long. If a player is running from second to third, what is her distance from home plate when she is 20 ft from third? [Swokowski]

6. The top part of a swimming pool is a rectangle of length 60 ft and width 30 ft. The depth of the pool varies uniformly from 4 ft to 9 ft through a horizontal distance of 40 ft and then is level for the remaining 20 ft. If the pool is being filled with water, find the number of gallons of water in the pool when the depth at the deep end is 4 ft. (1 gal is approximately 0.1337 ft³). [Swokowski]

7. A stone dropped into a still pond sends out a circular ripple whose radius increases at a constant rate of 3 ft/sec. What is the area enclosed by the ripple after 10 sec? [Anton]

8. Grain pouring from a chute at the rate of 8 ft³/min forms a conical pile

whose altitude is always twice its radius. What is the volume of grain when the pile is 6 ft high? [Anton]

9. A police helicopter is flying due south at 100 mi/hr, and at a constant altitude of ½ mi. Below, a car is traveling west on a highway at 75 mi/hr. At the moment the helicopter crosses over the highway the car is 2 mi east of the helicopter. What is the distance between the car and helicopter 30 minutes later? [Anton]

10. Coffee is poured at a uniform rate of 2 cm³/sec into a cup whose bowl is shaped like a truncated cone. If the upper and lower radii of the cup are 4 cm and 2 cm and the height of the cup is 6 cm, what is the volume of the coffee when the coffee is halfway up? [Anton]

Fig. 17.10

A Geometric Look at Volume-of-Solids-of-Revolution Problems

Calculus students have great difficulty with problems requiring them to find the volumes of solids of revolution. On a multiple-choice final examination problem asking students to set up the integral for a routine volume-of-solid-of-revolution problem, the average rate of correct responses over the past several semesters at this university has been 25 percent. Among college freshmen, the most commonly made errors include revolving the region about an axis other than the correct one, attempting to sketch the solid of revolution using an incorrect reflection of the region through the axis of revolution, and being unable to choose appropriately between the "disk" method and the "cylindrical shell" method because of an inadequate sketch (Mundy 1981).

Understanding the derivation of volume-of-solid-of-revolution techniques and then solving problems by applying these techniques in a meaningful way taps a varied set of spatial-geometric skills. Consider the following problem, typical of those found in all calculus textbooks:

> Sketch the region R bounded by the graphs of the given equations and find the volume of the solid generated by revolving R about the indicated axis:
> $$x = y^3, \ x^2 + y = 0; \text{ about the } x\text{-axis.}$$

To approach this problem successfully, the student must

- represent the information provided graphically and determine the region to be revolved;
- sketch or visualize the solid of revolution and choose the appropriate method ("disks or washers" or "cylindrical shells");
- sketch or visualize a cross section of the solid, revolve a "representative slice" to form a cross section, or place a representative shell correctly;
- express the representative volume in terms of the given relationships and finally set up an integral to compute the volume.

Only the very last of these steps actually involves a calculus concept. All the other steps require visualization and graphical representations that can be anticipated in the secondary school geometry curriculum.

Activities can be devised for geometry students that will ready them for the several phases of the volume-of-solids-of-revolution problems they will encounter later on. For example, in preparing students to sketch revolved solids, teachers can introduce exercises involving the reflection of regions in lines (see fig. 17.11).

Sketch the reflection of the given region in the given line.

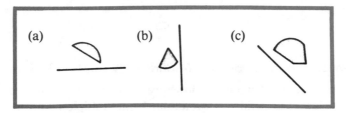

Fig. 17.11

The concept of a solid of revolution can be introduced in the context of the geometry course, ideally with the assistance of computer graphics that show the solid as it is swept out. An inexpensive substitute for such graphics are the crepe paper fold-out decorations available in card and gift shops. Wedding bells, Thanksgiving turkeys, Halloween witches' hats, and snowmen can be amazingly useful in demonstrating the concept of a solid of revolution. Asking students to sketch solids of revolution can follow the examples where the students are sketching reflections. Multiple choice problems like the one in figure 17.12 are an interesting means of approaching this phase.

Which of the following is the correct solid of revolution?

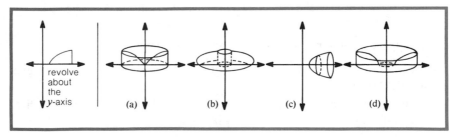

Fig. 17.12

Once students are comfortable with the concept of a solid of revolution, it is possible to combine their new understanding with their previous experience in geometry in finding volumes of solids. Examples such as the following anticipate the volume-of-solids-of revolution problems to be encountered in calculus but can be solved by the secondary school geometry student:

Sketch and find the volume of the solid formed by revolving about the x-axis the region bounded by $y = x$, $y = (4/3)x - 2$, and the x-axis.

Solution:
volume = "outer" volume − "inner" volume

$$= (\frac{1}{3})\pi 6^2 \cdot 6 - (\frac{1}{3})\pi 6^2 \cdot (4\frac{1}{2})$$

$$= 18\pi$$

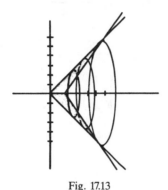

Fig. 17.13

Once the sketch is obtained (see fig. 17.13), it is clear that this volume can be computed using the formula for the volume of a cone; the volume of the solid is the volume of the outer cone minus the volume of the inner cone. The student does several of the same steps that are involved in a calculus volume-of-a-solid-of-revolution problem and must obtain an accurate visual representation of the problem before it is possible to obtain the volume using noncalculus techniques.

In calculus, the correct solution of volume-of-solids-of-revolution problems using "disks or washers" depends on the student's ability to visualize the representative cross section of the solid and to compute the volume of this representative cross section. Sketching the cross sections and labeling their dimensions in a solid-of-revolution problem is a noncalculus activity that can pose considerable problems for students. The final phases of the solution process also can be anticipated by appropriate exercises at the secondary level.

In figure 17.14 are several exercises that could be used at the secondary school geometry level to prepare students for the volume-of-solids-of-revolution problems they will encounter in calculus.

A Geometric Look at Calculus Problem Solving

Many physical applications of calculus involve definite integration. Two of the many applications discussed in any standard calculus course are determining the work performed or the force exerted by liquids. As with optimization and related-rates problems, students' difficulties with solving

Solids of Revolution

For each of the following regions and for each of the indicated axes, (a) sketch the solid of revolution, (b) find the volume of the solid of revolution, and (c) sketch a cross section of the solid of revolution and label its dimensions for a given x or y.

1.

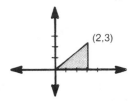

a. Revolve about X-axis
b. Revolve about Y-axis

2.

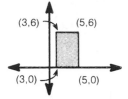

a. Revolve about X-axis
b Revolve about Y-axis

3.

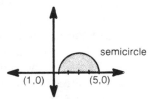

a. Revolve about X-axis
b. Revolve about Y = -6
(Requires volume of a torus formula)

4.

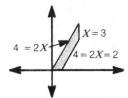

a. Revolve about X-axis
b Revolve about Y-axis

5.

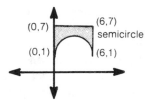

a. Revolve about X-axis
b. Revolve about Y-axis

6.

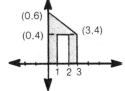

a. Revolve about X-axis
b. Revolve about Y-axis

7.

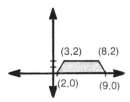

a. Revolve about X-axis
b. Revolve about X = 1

8.

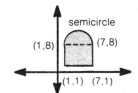

a. Revolve about Y-axis
b. Revolve about X = 1

Fig. 17.14

these types of applied problems do not usually stem from the difficulty of the calculus involved. Indeed, the actual integrals encountered are often very simple to evaluate. Rather, it is finding the appropriate integrand and limits of integration that poses the greatest difficulty for the student; this in turn depends on the accuracy and detail of the student's representation of the problem situation. The real "problem" is often one of geometry, not calculus. Consider an example:

> A sheet of metal in the shape of an equilateral triangle of side 3 feet is completely submerged vertically in oil having a density of 50 pounds per cubic foot. The triangle is positioned so that one side is parallel to and 2 feet below the surface of the oil and the opposite vertex points down (see fig. 17.15). What is the force exerted by the oil on one face of the triangle?

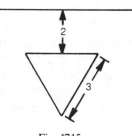

Fig. 17.15

In solving this problem, we must describe the width of the triangle as a function of depth and the range of depths the triangle occupies. In fact, the force exerted on one face of any plane region similarly submerged in liquid of density D pounds per cubic feet is given by the integral

$$D \int_a^b x \cdot w(x) \, dx$$

where x is the depth in feet, $w(x)$ is the width of the region at depth x, and a and b are the shallowest and deepest levels (in feet) respectively of the region.

Finding the function $w(x)$ is an exercise that requires the use of both the Pythagorean theorem and similar triangles. Stated specifically as a geometry problem, the problem above is written thus:

> In figure 17.16, $\triangle ABC$ is an equilateral triangle with sides of length 3. Segment DE is parallel to segment AB. P and Q are the midpoints of AB and DE respectively. If PQ has length h, find the length w of DE in terms of h.

Fig. 17.16

Since PC has length $3\sqrt{3}/2$ (by use of the Pythagorean theorem on right triangle BPC) and since $\triangle DEC$ and $\triangle ABC$ are similar, the following relationship holds for w and h:

$$\frac{3\sqrt{3}/2 - h}{w} = \frac{\sqrt{3}}{2}$$

Thus $w = 3 - (2\sqrt{3}/3)h$. Applying this result to the previous calculus problem, and noting that $h = (x - 2)$, yields

$$w(x) = 3 + (4\sqrt{3} - 2\sqrt{3}x)/3, \quad 2 \le x \le 2 + 3\sqrt{3}/2.$$

This problem points out a very different way of viewing geometric figures.

Instead of being simply a static set of points satisfying one or more properties, a geometric figure can be thought of as defining a dynamic functional
relationship between its dimensions. In a very real sense, the function $w(x)$
above "defines" the submerged triangle. We might even graph the function
$w(x)$ (see fig. 17.17).

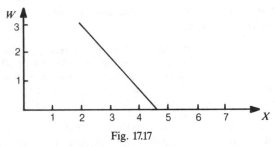

Fig. 17.17

It is interesting to note the different width functions that define different
plane regions submerged in a similar manner. For example, if the region is
a rectangle with one side parallel to the surface, then $w(x)$ is a constant
function over an appropriate interval. If, however, the region is a circle,
then the width function involves the square root of a quadratic expression
(see the illustrations in fig. 17.18).

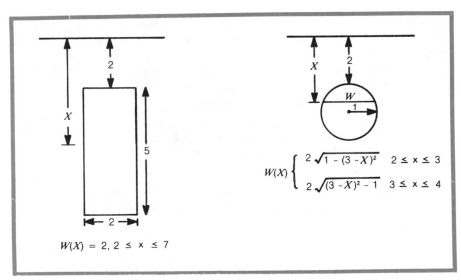

$$W(X)\begin{cases} 2\sqrt{1-(3-X)^2} & 2 \le x \le 3 \\ 2\sqrt{(3-X)^2-1} & 3 \le x \le 4 \end{cases}$$

$$W(X) = 2,\ 2 \le x \le 7$$

Fig. 17.18

In figure 17.19 are some exercises that enable high school students to focus
on the geometric aspects inherent in many applied calculus problems involving definite integrals.

Geometric Aspects of Definite Integration

1. Find the volume of a right circular cone inscribed in (or circumscribed about) a sphere of radius a as a function of its radius; its height.
2. Find the volume of a right circular cylinder inscribed in a sphere of radius a as a function of its radius; its height.

Find the "width function" that defines—

3. an isosceles triangle of height a and base b;
4. a scalene triangle with sides a, b, c.

Find the "cross-sectional area function" that defines—

5. a right circular cone of radius r and height h;
6. a hemisphere of radius r.

Fig. 17.19

Conclusion

Geometry has a central role to play in the high school curriculum, particularly the formal geometry course with its emphasis on proof. In this article, it is suggested that there are other roles for geometry—it is especially suited for providing calculus readiness. The ideas presented are only a beginning, and any standard calculus textbook is a source for additional problems. These suggestions may be integrated into an existing tenth-grade geometry course or used as part of a half-year, senior-level course in mathematical problem solving.

The approach suggested in this article may be more useful for most students in high school than an introduction to polynomial calculus.

BIBLIOGRAPHY

Anton, Howard. *Calculus with Analytic Geometry.* New York: John Wiley & Sons, 1984.

Lean, G., and Ken Clements. "Spatial Ability, Visual Imagery, and Mathematical Performance." *Educational Studies in Mathematics* 12 (1981): 267–99.

Meserve, Bruce. "Geometry as a Gateway to Mathematics." In *Developments in Mathematical Education: Proceedings of the Second International Congress on Mathematical Education,* edited by A. G. Howson. Cambridge: Cambridge University Press, 1973.

Mundy, Joan F. "Spatial Ability, Mathematics Achievement, and Spatial Training in Male and Female Calculus Students." (Doctoral dissertation, University of New Hampshire, 1980). *Dissertation Abstracts International* 41 (1981): 4633A (University Microfilms No. 8108871).

Purcell, Edwin J. *Calculus with Analytic Geometry.* Englewood Cliffs, N.J.: Prentice-Hall, 1978.

Swokowski, Earl W. *Calculus with Analytic Geometry.* Boston: Prindle, Weber & Schmidt, 1984.

18

Geometric Counting Problems

Bonnie H. Litwiller
David R. Duncan

MANY secondary mathematics students have the impression that algebra and geometry are totally distinct one from the other. There is, of course, a partial merging of the concepts of these two branches of mathematics when analytic geometry is considered. We shall present still another union of algebra and geometry, namely, geometric counting problems.

The basic setting for geometric counting problems involves a grid or lattice. Geometric figures of selected types are drawn on this grid. The problem is to count the total number of such figures that can be drawn. Such a problem can be approached at two levels. First, an informal exploration can lead to a conjecture concerning the number of figures it is possible to draw. Then, an algebraic proof can follow, depending, of course, on the mathematical maturity of the student. We shall illustrate five such problems.

Problem 1

How many squares can be drawn on an $n \times n$ checkerboard? When students are first asked this question for a regular checkerboard, a common response is "Sixty-four squares," since a checkerboard is usually visualized as a board containing eight rows and eight columns of alternately red and black squares.

A more careful examination of this problem, however, reveals that another interpretation is possible; namely, to find all possible squares of all sizes using only the segments indicated on the board. For instance, figure 18.1 denotes some of the possible squares that can be found on a checkerboard.

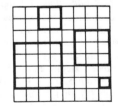

Fig. 18.1

To analyze the situation more closely, draw 1×1, 2×2, and 3×3 "checkerboards" and count the number of squares that it is possible to draw for each. Clearly, a 1×1 checkerboard contains only one square. On a 2×2 checkerboard, there are four 1×1 squares and one 2×2 square. Thus there are five squares on a 2×2 checkerboard.

On a 3×3 checkerboard (fig. 18.2) there are nine 1×1 squares, and

210

one 3×3 square. In addition, there are four 2×2 squares, as indicated by the shading. Thus there are fourteen squares on a 3×3 checkerboard. Table 18.1 reports the results of the problem, including a 4×4 checkerboard.

Fig. 18.2

Table 18.1

Size of Checkerboard	No. of 1×1 Squares	No. of 2×2 Squares	No. of 3×3 Squares	No. of 4×4 Squares	Total
1×1	1				1
2×2	4	1			5
3×3	9	4	1		14
4×4	16	9	4	1	30

The data gave rise to the following conjecture: *The number of squares on an $n \times n$ checkerboard is* $1^2 + 2^2 + 3^2 + \cdots + n^2$. Call this Formula 1.

Now we shall prove that the conjecture is correct, using mathematical induction.

Proof of Formula 1: The number of squares on an $n \times n$ checkerboard is $1^2 + 2^2 + 3^2 + \cdots + n^2$.

Step 1. Consider the case when $n = 1$. On a 1×1 checkerboard there is clearly only one square. Since $1 = 1^2$, Formula 1 is true when $n = 1$.

Step 2. Assume that Formula 1 is true when n is replaced with some positive integer k. That is, assume that a $k \times k$ checkerboard contains $1^2 + 2^2 + 3^2 + \cdots + k^2$ squares.

Step 3. If n is replaced with $(k + 1)$, prove that Formula 1 still holds. That is, prove that a $(k + 1) \times (k + 1)$ checkerboard contains $1^2 + 2^2 + 3^2 + \cdots + (k + 1)^2$ squares.

Figure 18.3 displays a $(k + 1) \times (k + 1)$ checkerboard. The checkerboard consists of two parts—an "original" $k \times k$ checkerboard and a "border" whose squares are numbered along the top and right-hand side.

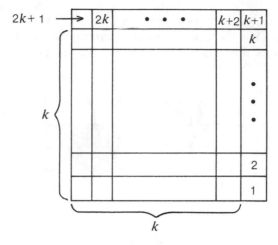

Fig. 18.3

The squares on figure 18.3 may be counted in four categories:

Category A

The squares completely contained in the original part. By step 2 there are $1^2 + 2^2 + 3^2 + \cdots + k^2$ squares on this part of the checkerboard.

Category B

Those squares intersecting the border of squares $k + 2$ through $2k + 1$ on the top of the figure but not intersecting the border of squares 1 through k on the right-hand side of the figure. None of those squares would contain the upper right-hand corner square $k + 1$. Count the squares of this category as follows:

1. There is one $k \times k$ square.
2. There are two $(k - 1) \times (k - 1)$ squares.
3. There are three $(k - 2) \times (k - 2)$ squares. (Have your students draw them.)

   .
   .
   .

k. There are k unit squares. These are simply the squares in the top border not including the upper right-hand square.

The total number of squares in category B is thus $1 + 2 + 3 + \cdots + k$.

Category C

Those squares intersecting the border of squares 1 through k on the right of figure 18.3 but not intersecting the border of squares $k + 2$ through $2k + 1$ on the top of the figure. None of these squares would contain the upper right-hand corner square $k + 1$. An argument identical to that of category B implies that the total number of squares in category C is $1 + 2 + 3 + \cdots + k$.

Category D

Those squares containing the upper right-hand corner square $k + 1$. There are exactly $(k + 1)$ squares in this category—one each of sizes 1×1, 2×2, 3×3, $\cdots$, $(k + 1) \times (k + 1)$.

The total number of squares in figure 18.3 is computed by adding the results of categories A through D. This sum can be expressed as follows:

$$\underbrace{[1^2 + 2^2 + 3^2 + \cdots + k^2]}_{A} + \underbrace{2[1 + 2 + 3 + \cdots + k]}_{B \text{ and } C} + \underbrace{[k + 1]}_{D}$$

$$= [1^2 + 2^2 + 3^2 + \cdots + k^2] + 2\,\frac{k(k + 1)}{2} + [k + 1]$$

$$= [1^2 + 2^2 + 3^2 + \cdots + k^2] + (k + 1)(k + 1) \text{ (by factoring)}$$

$$= (1^2 + 2^2 + 3^2 + \cdots + k^2) + (k + 1)^2$$

$$= [1^2 + 2^2 + 3^2 + \cdots + (k + 1)^2].$$

Step 3 is thus proved and the induction is complete. Our conjecture is therefore established.

Problem 2

How many rectangles can be drawn on an $n \times n$ checkerboard? Recall that a square is a special case of a rectangle.

To find a pattern, consider small checkerboards of varying sizes. For a 1×1 checkerboard there exists only one rectangle—just the single square itself. For a 2×2 checkerboard there are 9 rectangles:

4—1×1 squares
1—2×2 square
2—1×2 rectangles
2—2×1 rectangles
9 rectangles

Note that a rectangle of dimension $a \times b$ will be considered to have horizontal length a and vertical length b. By symmetry, the number of $a \times b$ rectangles is the same as the number of $b \times a$ rectangles.

For a 3×3 checkerboard there are 36 rectangles:

9—1×1 squares
4—2×2 squares
1—3×3 square
6—1×2 rectangles
6—2×1 rectangles
3—1×3 rectangles
3—3×1 rectangles
2—2×3 rectangles
2—3×2 rectangles
36 rectangles

The reader can ascertain that the 4×4 checkerboard contains 100 rectangles and the 5×5 checkerboard contains 225 rectangles. From the prior sequence of results $(1, 9, 36, 100, 225, \cdots)$, one can speculate that the number of rectangles that can be found on an $n \times n$ checkerboard is $1^3 + 2^3 + \cdots + n^3$, or equivalently $(1 + 2 + \cdots + n)^2$. Call this Formula 2. Note that this last formula is the square nth triangular number.

The general proof that the number of rectangles contained on a checkerboard is $(1 + 2 + \cdots + n)^2$ follows by mathematical induction; the proof is somewhat similar to that of problem 1 and thus is not included here in the interest of space.

Problem 3

How many cubes are in an $n \times n \times n$ cube? For example, consider a large block of cube sugar consisting of n^3 small sugar cubes. How many cubes of all sizes does it contain?

When a square checkerboard was considered, Formula 1 involved the sum of squares. When a cube is used, one possible conjecture would be that Formula 1 could be modified by replacing the squares by cubes.

CONJECTURE. *The total number of cubes (of all sizes) in an $n \times n \times n$ cube is* $1^3 + 2^3 + 3^3 + \cdots + n^3$. Call this Formula 3.

The second task would be to check some simple cases to verify the reasonableness of the conjecture and then to prove that Formula 3 is correct. The proof, by mathematical induction, contains similarities both to the proof for Formula 1 and the proof for Formula 4, which follows.

Problem 4

How many rectangular solids are in an $n \times n \times n$ cube?

When a square checkerboard was considered, Formula 2 involved the square of a sum. When a cube is considered, one possible conjecture would be that Formula 2 could be modified by replacing the square with a cube.

CONJECTURE. *The total number of rectangular solids in an $n \times n \times n$ cube is* $(1 + 2 + 3 + \cdots + n)$.[3] Call this Formula 4.

The following proof of Formula 4 employs mathematical induction.

Proof of Formula 4: The number of rectangular solids in an $n \times n \times n$ cube is $(1 + 2 + 3 + \cdots + n)$.

Step 1. Consider the case when $n = 1$. In a $1 \times 1 \times 1$ cube there is clearly one rectangular solid, the $1 \times 1 \times 1$ cube itself. Since $1 = 1^3$, Formula 4 is true when $n = 1$.

Step 2. Assume that Formula 4 is true when n is replaced by some positive integer k. That is, assume that a $k \times k \times k$ cube contains $(1 + 2 + 3 + \cdots + k)^3$ rectangular solids.

Step 3. If n is replaced with $(k + 1)$, prove that Formula 4 still holds. That is, prove that a $(k + 1) \times (k + 1) \times (k + 1)$ cube contains $[1 + 2 + 3 + \cdots + (k + 1)]^3$ rectangular solids.

Figure 18.4 displays a $(k + 1) \times (k + 1) \times (k + 1)$ cube. The cube consists of two parts—an "original" $k \times k \times k$ cube (which is totally hidden behind the visible faces) and three additional $(k + 1) \times (k + 1) \times (1)$ "border faces," which are labeled face P, face Q, and face R. These border faces intersect in pairs along three edges as shown. For instance, faces P and Q intersect in a vertical tower of $(k + 1)$ cubes that are $1 \times 1 \times 1$, k of which have one face-P surface and one face-Q surface, and one of which has one surface on each of faces P, Q, and R.

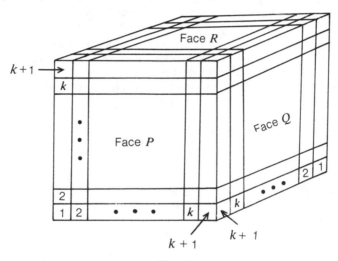

Fig. 18.4

The rectangular solids in figure 18.4 may be counted in eight categories:

Category A

The rectangular solids completely contained in the original $k \times k \times k$ cube. None of these rectangular solids intersects any of faces P, Q, or R. By step 2 there are $(1 + 2 + 3 + \cdots + k)^3$ rectangular solids in this part of the cube.

Category B

Those rectangular solids intersecting face P but not intersecting either of faces Q or R. Each of these rectangular solids has one face-P surface and intersects face P of figure 18.4 somewhere within the lower left $k \times k$ portion of that border face.

Conversely, each rectangle on that portion of border face P yields $(k + 1)$ rectangular solids of depths 1, 2, 3, $\cdots$, $k + 1$. By Formula 2 there are exactly $(1 + 2 + 3 + \cdots + k)^2$ such rectangles on the lower left portion of that border face. Hence, there are exactly $(1 + 2 + 3 + \cdots + k)^2 \times (k + 1)$ rectangular solids in category B.

Category C

Those rectangular solids intersecting face Q but not intersecting either of faces P or R. The same reasoning as that of category B implies that there are $(1 + 2 + 3 + \cdots + k)^2 \times (k + 1)$ rectangular solids in category C.

Category D

Those rectangular solids intersecting face R but not intersecting either of faces P or Q. Again there are $(1 + 2 + 3 + \cdots + k)^2 \times (k + 1)$ rectangular solids in category D.

Category E

Those rectangular solids containing one surface on face P and one surface on border face Q but no surface on face R. Count the rectangular solids of this category as follows:

1. There are $1 \times (k + 1) \times (k + 1)$ rectangular solids of height k. There is only one possible vertical placement for such a rectangular solid; there are $(k + 1)$ choices for each of its other two dimensions.

2. There are $2 \times (k + 1) \times (k + 1)$ rectangular solids of height $(k - 1)$. There are two possible vertical positions; there are $(k + 1)$ choices for each of the other two dimensions.

3. There are $3 \times (k + 1) \times (k + 1)$ rectangular solids of height $(k - 2)$.

$$\vdots$$

k. There are $k \times (k + 1) \times (k + 1)$ rectangular solids of height 1. The total number of rectangular solids in this category is thus $(1 + 2 + 3 + \cdots + k) \times (k + 1)^2$.

Category F

Those rectangular solids containing one surface on each of faces P and R but none on face Q. The same reasoning as that of category E implies that there are $(1 + 2 + 3 + \cdots + k) \times (k + 1)^2$ rectangular solids in this category.

Category G

Those rectangular solids containing one surface on each of faces Q and R but none on face P. Again there are $(1 + 2 + 3 + \cdots + k) \times (k + 1)^2$ rectangular solids in this category.

Category H

Those rectangular solids containing surfaces on each of faces P, Q, and R. To form such a rectangular solid, there are $(k + 1)$ possibilities for each of its three dimensions. Consequently, there are $(k + 1)^3$ such rectangular solids in this category.

The total number of rectangular solids contained in figure 18.4 is computed by adding the number of rectangular solids in categories A through H. This sum is

$$\underbrace{[(1 + 2 + \cdots + k)^3]}_{A} + \underbrace{3[(1 + 2 + \cdots + k)^2(k + 1)]}_{B, C, \text{ and } D}$$

$$+ \underbrace{3[(1 + 2 + \cdots + k)(k + 2)^2]}_{E, F, \text{ and } G} + \underbrace{[(k + 1)^3]}_{H}$$

$$= [(1 + 2 + \cdots + k) + (k + 1)]^3 \text{ (by factoring)}$$

$$= [1 + 2 + \cdots + (k + 1)]^3.$$

Step 3 is thus proved and the induction is complete.

An interesting pattern is formed by the four formulas resulting from the first four problems.

1. For a square of side n, $\displaystyle\sum_{k=1}^{n} k^2$ squares can be found.

2. For a square of side n, $\displaystyle\sum_{k=1}^{n} k^3$ rectangles can be found.

3. For a cube of side n, $\left[\displaystyle\sum_{k=1}^{n} k\right]^2$ cubes can be found.

4. For a cube of side n, $\left[\displaystyle\sum_{k=1}^{n} k\right]^3$ rectangular solids can be found.

Formulas 2 and 3 yield the same number for a given value of n. This is surprising, since Formulas 2 and 3 are generated from two quite different problems. There is no a priori reason to suspect that the number of rectangles in an $n \times n$ square is the same as the number of cubes in an $n \times n \times n$ cube.

Conjectures 2 and 4 dealt with rectangular shapes on two- and three-dimensional figures. A one-dimensional figure, a line segment calibrated at integer points, could also be considered. It is easy to deduce that there are $(1 + 2 + 3 + \cdots + n)^1$ line segments on a calibrated line segment of length n. Table 18.2 reports results for one-, two-, and three-dimensional figures. What would be comparable problems in spaces of four or higher dimension? Would the pattern of table 18.2 continue for these dimensions? How could such results be pictured and proved?

TABLE 18.2

Dimension	No. of "Rectangular Geometric Shapes" on a Figure n Units per Side
1	$(1 + 2 + 3 + \cdots + n)^1$
2	$(1 + 2 + 3 + \cdots + n)^2$
3	$(1 + 2 + 3 + \cdots + n)^3$

Problem 5

How many parallelograms can be drawn on an $n \times n$ checkerboard? This problem is different from the preceding five in that although the vertices are points of the original lattice, the line segments connecting them are not necessarily part of the original grid.

To organize the counting process, first identify all the rectangles with horizontal and vertical sides on the checkerboard. For each such rectangle, all the parallelograms are counted so that—

1. these parallelograms are contained in the rectangle being considered;
2. no smaller rectangle contains these parallelograms.

The parallelograms described in 1 and 2 will be called the "maximal parallelograms" for the given rectangle. The parallelograms thus contained in each such rectangle are counted, and their total gives the number of parallelograms contained on the entire checkerboard.

The following five cases organize the maximal parallelograms contained in an arbitrary $a \times b$ rectangle of horizontal and vertical sides. Here and henceforth, a denotes the horizontal dimension and b the vertical dimension of the rectangle.

Case I. The $a \times b$ rectangle itself. Only one rectangle of horizontal and vertical sides, the $a \times b$ rectangle itself, is a maximal parallelogram in an $a \times b$ rectangle.

Case II. Those nonrectangular parallelograms that have two horizontal sides. Clearly one pair of opposite vertices of such a parallelogram must coincide with a pair of opposite vertices of the rectangle; otherwise, the parallelogram would not be maximal in the rectangle. These cases are illustrated in figure 18.5.

Maximal parallelogram

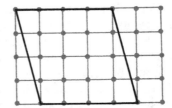

Nonmaximal parallelogram

Fig. 18.5

If such a parallelogram contains the upper left and lower right vertices of the rectangle, it is completely determined by the length of a horizontal side. Since $(a - 1)$ such lengths are possible, there are $(a - 1)$ such parallelograms in this upper left to lower right orientation. Also, another $(a - 1)$ maximal parallelograms in this case will contain the upper right and lower left vertices of the rectangle. Thus, the total number of parallelograms in this case is $2(a - 1)$.

Case III. Those nonrectangular parallelograms that have two vertical sides. The counting process is symmetric to Case II; hence, there are $2(b - 1)$ such parallelograms.

Case IV. Those parallelograms that have neither horizontal nor vertical sides but that have their four vertices on the four sides of the rectangle. Two independent choices are necessary in order to determine such a parallelogram. One such choice locates the vertex of the parallelogram that is continued on the upper horizontal base of the rectangle; the second choice locates the point on the right vertical side of the rectangle. After these two choices are made, only one parallelogram can be drawn that has its other two vertices on the other two sides of the rectangle.

There are $(a + 1)$ lattice points on the upper horizontal base of the rectangle. Since the four vertices of the parallelogram must lie on the four sides of the rectangle, the vertices of the rectangle itself cannot be used as vertices of the parallelogram. There are, therefore, $(a - 1)$ lattice points on the upper horizontal base available for selection as a vertex of the parallelogram. Similarly, there are $(b - 1)$ lattice points on the right vertical side of the rectangle available for selection as a vertex of the parallelogram. Consequently, the parallelograms of this case number $(a - 1)(b - 1)$.

Case V. Those parallelograms that have—
1. neither horizontal nor vertical sides;
2. exactly one pair of opposite vertices coincident with a pair of opposite vertices of the rectangle;
3. the other pair of opposite vertices lying in the interior of the rectangle.

Consider first the situation in which the upper left and lower right vertices of the rectangle are coincident with the vertices of the parallelogram. The other two vertices of the parallelogram must then lie on opposite sides of the upper left to lower right rectangle diagonal. This case is shown in figure 18.6. Observe that if one of the internal vertices of the parallelogram is chosen, the fourth vertex of the parallelogram is uniquely determined. How many choices are available for the first interior vertex to be chosen?

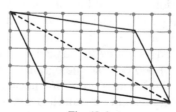

Fig. 18.6

The interior of the $a \times b$ rectangle contains $(a - 1) \cdot (b - 1)$ lattice points. Any lattice point on the upper left to lower right diagonal cannot be used, lest a degenerate parallelogram result. There are [G.C.D.$(a, b) - 1$] such interior lattice points on that diagonal. (*Note:* "G.C.D.(a, b)" denotes the greatest common divisor of a and b.) Consequently, there are

$$\frac{(a - b)(b - 1) - \text{G.C.D.}(a, b) + 1}{2}$$

such lattice points in the interior of the rectangle on one side of the upper left to lower right rectangle diagonal available for selection as the first interior point for the vertex of the parallelogram. This is therefore the number of parallelograms in this case that contain the upper left and lower right vertices of the rectangle. Since the parallelogram in this case could instead contain the upper right and lower left vertices of the rectangle, there are twice the number of parallelograms as just counted, or

$$(a - 1)(b - 1) - \text{G.C.D.}(a, b) + 1.$$

The total number of maximal parallelograms in a rectangle of dimensions $a \times b$ is

$$1 + 2(a - 1) + 2(b - 1) + (a - 1)(b - 1) + [(a - 1)(b - 1) \\ - \text{G.C.D.}(a, b) + 1].$$

How many parallelograms are contained on a checkerboard of dimensions 8×8? For each rectangle having horizontal and vertical sides in this checkerboard, count the number of maximal parallelograms it contains. For example, consider a rectangle of dimensions 2×4 on the checkerboard. This rectangle contains

$$1 + 2(2 - 1) + 2(4 - 1) + (2 - 1)(4 - 1) + [(2 - 1)(4 - 1) \\ - \text{G.C.D.}(2, 4) + 1] = 1 + 2 + 6 + 3 + 2, \text{ or } 14, \text{ parallelograms.}$$

However, there is not just one 2×4 rectangle on the 8×8 checkerboard. There are, in fact, seven horizontal positions and five vertical positions that a 2×4 rectangle can assume, or $7(5) = 35$ such rectangles. (In general the number of $a \times b$ rectangles on an 8×8 checkerboard is $(8 - a + 1)(8 - b + 1) = (9 - a)(9 - b)$.) Since each of these 35 rectangles contains 14 maximal parallelograms, there are $35(14) = 490$ maximal parallelograms in rectangles of dimension 2×4. When all such cases have been counted, exactly 26 904 maximal parallelograms can be identified.

How rapidly does the number of parallelograms increase as the dimensions of the checkerboard increase? In particular, how large a checkerboard is necessary to achieve 1 000 000 parallelograms? We find that a 15×15 checkerboard yields 900 052 and a 16×16 checkerboard yields 1 299 147 parallelograms. How large a checkerboard do you predict would yield a billion parallelograms?

Many additional questions of these types may be asked:

1. How many parallelograms can be drawn on an $n \times n$ rhombus?

2. How many equilateral triangles can be drawn on the regular hexagon shown in figure 18.7?

3. How many oblique squares can be drawn on a checkerboard? (See fig. 18.8.)

4. How many triangles can be drawn on an $n \times n \times n$ equilateral triangular array? (See fig. 18.9.)

Fig. 18.7

Fig. 18.8

Fig. 18.9

In each of these and other problems that you and your students might develop, the problem should first be investigated for simple special cases, a general conjecture should be made (induction), and a deductive proof can be considered (if your students are at the appropriate level of mathematical maturity).

BIBLIOGRAPHY

Duncan, David R., and Bonnie H. Litwiller. "Checkerboards and Sugar Cubes: Geometric Counting Patterns." *Two-Year College Mathematics Journal* 1(Spring 1973): 24–27.

————. "Triangles and Triangular Numbers: A Geometric Counting Problem." *Pentagon* 34 (Spring 1975): 84–92.

————. "Parallelograms on an $n \times n$ Rhombus: A Geometric Counting Problem." *Mathematics in School* 8 (November 1979): 26–29.

————. "Counting Squares and Cubes." *Australian Mathematics Teacher* 38 (July 1982): 2–4.

Litwiller, Bonnie H., and David R. Duncan. "Counting the Oblique Squares on a Checkerboard." *New England Mathematics Journal* 11 (May 1979): 13–16.

————. "Parallelograms by the Million: A Geometric Counting Problem." *Australian Mathematics Teacher* 35 (July 1979): 13–15.

————. "Rectangles, Diagonals, and Lattice Points: An Application of the GCD." *Ohio Journal of School Mathematics* 3 (November 1979): 9–11.

————. "Counting Rectangles and Rectangular Solids: Proof by Induction." *New York State Mathematics Teachers' Journal* 33 (1983): 41–45.

19

Activities with Teachers Based on Cognitive Research

Rina Hershkowitz
Maxim Bruckheimer
Shlomo Vinner

$\mathbf{A}$ BASIC knowledge of geometry is fundamental for children to interact effectively with their environment as well as for them to enter a more formal study of geometry. This basic knowledge, which comprises geometric concepts, their attributes, and simple relationships, should, in general, be acquired through geometrical experiences prior to secondary school.

If students are to learn these fundamentals, then it is important for elementary school teachers to be comfortable with these ideas and the ways to help children learn. But research (Hershkowitz and Vinner 1984) has shown that teachers have patterns of misconceptions similar to those of students in grades 5–8.

After describing some patterns of similarity between students' and teachers' misconceptions, we shall discuss activities for teachers that can not only help them learn geometric concepts but also provide a useful model for them to use in the classroom.

Examples of Geometric Concept Images
Held by Students and Teachers

The notion of concept image was introduced as the collection of mental images that an individual has of a given concept (Vinner and Hershkowitz 1980). An individual's concept image may be complete, partial, or incorrect. A *partial* concept image does not contain all the aspects included in the concept definition. An *incorrect* concept image includes items that do not belong.

The examples that follow were administered to 518 students (grades 5–8), 142 preservice elementary teachers (PRE) and 25 in-service elementary teachers (ST) in Israel. The responses to these items allowed us to compare student and teacher concept images of certain geometrical concepts and some possible factors that influence their formation.

Example 1: The Angle Concept

If children understand the concept of an angle, then they realize that the drawing of an angle on a sheet of paper represents only part of that angle. This understanding was assessed by the item in figure 19.1. The results suggest that less than half of students conceive of an angle as an "infinite entity" (from 25% in the fifth grade to 50% in the eighth grade). Similarly, only slightly more than half of the teachers (68% of the prospective teachers and 55% of the in-service teachers) had the proper concept of an angle.

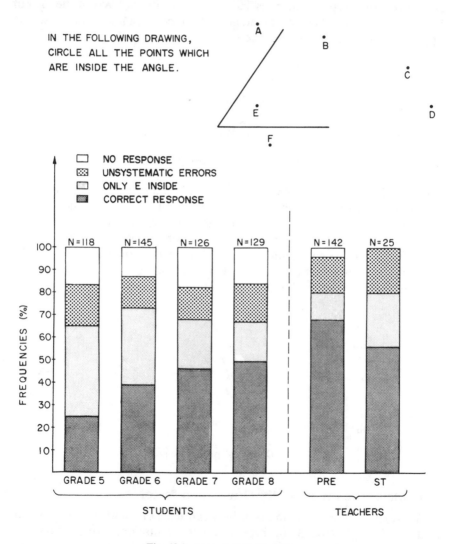

Fig. 19.1. Angle task and results

Examples 2 and 3: Altitude of a Triangle and Diagonals of a Polygon

On tasks that required drawing an altitude in different types of triangles or all the diagonals from one vertex of concave polygons, the teachers (ST and PRE) performed only a little better than the students. Many teachers, like many students, have incomplete concept images or concept images that include incorrect elements. Here are two examples.

a) When teachers and students were asked to draw the altitude to side *a* in the obtuse-angled triangle (fig. 19.2), some drew the median to *a* and some drew the perpendicular bisector of *a*. They did not accept the idea of an exterior altitude. The incompleteness of this concept image and its wrong application were expressed in exactly the same way by the students.

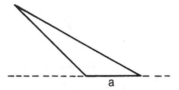

Fig. 19.2. Obtuse-angled triangle

b) Teachers, as well as students, drew only interior diagonals from the vertex *A* of concave polygons (see fig. 19.3). In the case of the concave quadrilateral (fig. 19.3c), the "inside" notion of their concept image of a diagonal caused confusion. Rejecting the possibility of an "exterior" diagonal, they either drew nothing at all (40% of the students; 12% of the teachers) or drew something like the dotted line in figure 19.3c (53% of the students; 49% of the teachers).

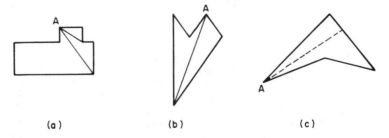

(a)   (b)   (c)

Fig. 19.3. "Diagonals" of concave polygons

Example 4: The Impact of the Orientation Factor

Figure 19.4 shows a comparison between teachers' and students' identification of a right triangle in three different orientations. (In the task they had to identify all the right triangles in a given collection of triangles.)

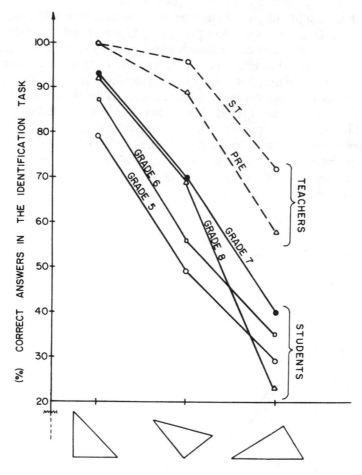

Fig. 19.4. Success of students and teachers in the identification of right triangles

Although, not surprisingly, teachers did better, their responses again exhibit the same pattern as those of the students. Thus their success was greatest for the right triangle in the upright position as usually drawn, decreased when the triangle was rotated through about 45°, and decreased drastically when the right angle was "at the top."

Example 5: The "Bitrian"

The purpose of this item was to investigate the role of a verbal definition in the formation of the relevant concept. We devised the following definition:

A "bitrian" is a geometric shape consisting of two triangles having a common vertex. (One point serves as a vertex of both triangles.)

One half of each of our groups (students and teachers) was asked to identify bitrians among other shapes, and the other half was asked to construct two bitrians. The frequencies of the shapes constructed by students and teachers (PRE and ST) are shown in figure 19.5. The pattern for teachers and students was very similar in both bitrian tasks (Hershkowitz and Vinner 1984). In both tasks the teachers and students started from the same point: the concept was new to all of them. The verbal definition created very similar concept images in both populations. Thus the relative frequency of concept examples showed the same pattern for both populations. In terms of mathematical validity there is nothing to distinguish bitrian (i) in figure 19.5 from bitrian (v), for example, but there would seem to be a considerable psychological difference.

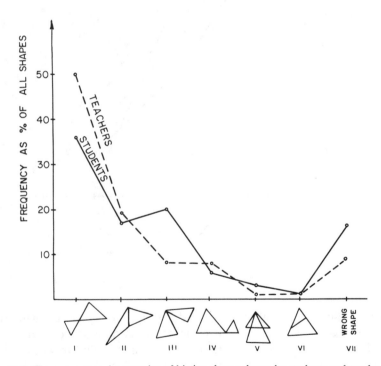

Fig. 19.5. The percentage frequencies of bitrian shapes drawn by students and teachers

The examples above, which are part of a much larger study, give evidence of the low level of knowledge concerning basic geometrical figures and their attributes that students have at the end of elementary school and the beginning of middle school. Van Hiele–based studies (Usiskin 1982; Hoffer 1983) showed that this causes many difficulties in upper levels of geometrical learning. The low level of geometrical knowledge that was also found in preservice and in-service elementary school teachers in Israel raises doubts about their ability to change this situation. There is an obvious need for preservice and in-service teachers to participate in activities dealing with basic geometrical concepts and their attributes.

A more critical point is the similarity evidenced between teachers and students in the patterns of incomplete or incorrect concept images. This raises the conjecture that the processes of forming geometrical concepts and the factors that inhibit this formation act similarly on individuals— students, student teachers, and teachers. It seems that there is a need to make the teacher, or the future teacher, familar with these processes and their associated misconceptions. These conclusions served as the basis for the following activities.

Activities for Teachers

In planning all our "remedial" in-service work with teachers, we take the view that teaching the teachers in the same way they were previously taught would be ineffective, boring, and insulting. It seems reasonable to suppose that if teachers are made explicitly aware of incorrect or incomplete concept images, they will be in a better position to understand the causes of student errors and misconceptions and hence to give appropriate instruction in concept formation in the classroom. Therefore, we designed activities that allow teachers to acquire concepts as well as an understanding of the processes and difficulties involved in concept formation. The explicit goals of the activities were to—

1. improve the teachers' concept images of some geometric concepts;
2. develop the teachers' understanding of
   - the role of concept definition;
   - the role of concept examples and relevant nonexamples;
   - the role of critical and noncritical attributes;[1]
3. model appropriate classroom teaching strategies;
4. provide experience in evaluating student difficulties and misconceptions.

We include two sample activities and the typical reactions of one group of twenty elementary teachers who attended a mathematics workshop one day each week throughout the school year.

---

1. *Critical attributes* are those attributes that an example must have in order to be an example of the concept. *Noncritical attributes* are those attributes that only particular examples have.

Activity 1: The "Trianquad"

Stage 1. An unstructured discussion is held on common ways of teaching basic geometrical concepts from early childhood, with special focus on the role of definition, concept examples, and ways of representing examples in the classroom and in textbooks.

Stage 2. The "trianquad" exercise is given to the participants (see fig. 19.6) as a possible teaching strategy (Herron et al. 1976). This task presents the formation of a concept through successive trials, using a set of examples and nonexamples. The participant gradually discovers what attributes a trianquad has and does not have. At the end, the definition of a trianquad is required.

Seventeen of the twenty teachers gave a definition that included the three

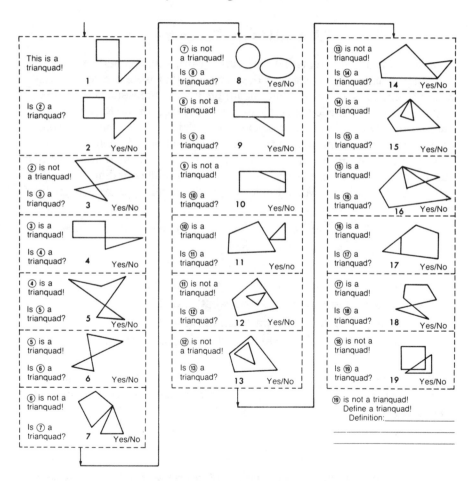

Fig. 19.6. The trianquad exercise

critical attributes of the trianquad (a geometrical figure consisting of a quadrilateral and a triangle with a common vertex). The remaining three teachers had mistakes in one of the attributes.

Four of the seventeen teachers added unnecessary or incorrect information based on noncritical attributes of particular examples. Here are some examples:

- The figures can be contained in one another, overlap, or be separated from one another. (Unnecessary)
- The triangle is "smaller" than the quadrilateral. (Incorrect)
- Trianquad is a seven-sided polygon. (Incorrect)

It is worth noting that the teachers' ability to verbalize the definition was much better than that of grade 6 and grade 8 students on a similar activity.

Stage 3. Analysis of the trianquad exercise is done by guided discussion. The following are excerpts from this discussion (*R* denotes the researcher and *T* denotes the teachers).

R: How is the concept of a trianquad conveyed in this activity?

T: Through examples.

T: Some of them were mistakes—the examples were not true.

T: "They" indicated our errors.

T: There was feedback too . . .

The discussion continued to the definition of a trianquad. We discussed why "seven-sided polygon" was an incorrect definition (we do not have seven sides in example 10 and yet it is a trianquad), and so on.

T: What about the common vertex; does it have to be only one?

R: What is your opinion?

Several teachers: Yes. No. I wrote only one. No. I wrote at least one . . .

T: "They" did not give us an example of a trianquad that has more than one common vertex, like this ⬜▷ , so it is only one.

T: Yes, but on the other hand, "they" did not give us an example with two common vertices and tell us it is not a trianquad, so we don't know.

Thus the discussion led naturally to the characteristics of a concept definition. We agreed that such a definition should be minimal on the one hand but unambiguous on the other. In another part of the discussion, the subject of characteristics of examples and nonexamples was raised.

R: For what items was your guess wrong and why?

T: I made a wrong guess for number 5 ◁ because it is not a "regular" quadrilateral.

T: I was wrong in number 7 ◇ because I had the

impression that the sides must lie on a straight line.

T: The examples there teach us, but on the other hand we should be careful.

T: Number 10 ▭ was not clear. I was not wrong there, but it was hard for me since I saw it as a pentagon and a triangle with a common side.

We see that teachers noticed the misleading effect that some example attributes can have (such as the noncritical attributes in no. 7), and the perceptual factors (as in no. 10), that inhibit our ability to recognize some examples of the concept.

Stage 4. We used of the trianquad exercise as a model of concept acquisition. We chose concepts whose images were not well developed for the student and teacher population. One example involved drawing the altitude to the side in a given triangle (see fig. 19.7).

At the beginning the researcher and the teachers analyzed together the concept of the altitude of a triangle, and then the researcher presented a few typical student mistakes and discussed them with the teachers. Figure 19.8 shows the response of one student.

R: What about this student? What did he draw?

T: He succeeded in the isosceles triangle [no. i] and in the second one [no. ii].

T: In the obtuse triangle [no. iii] he decided that the vertex is at the bottom.

T: This was easier for him.

T: Here the required altitude ought to be to the extension of side *a*. This business of an outside altitude is not clear to the student.

It is interesting to note that eleven of the twenty teachers in this group did not know how to draw an altitude to side *a* in the obtuse triangle. Five of them made the same mistake as the student in figure 19.8.

After examining and discussing together a few more student mistakes, the teachers were asked to work in groups of two or three on a collection of most frequent student errors on the altitude task. They were asked to analyze the response of each student, explain the answers, and order each, from worst to correct. In response, a few of the teachers created a hierarchy of answers ranging from those without a single critical attribute of the altitude to those that had at least one and so on, to the correct drawing. For most of the teachers, this exercise proved too difficult and would seem to need modification.

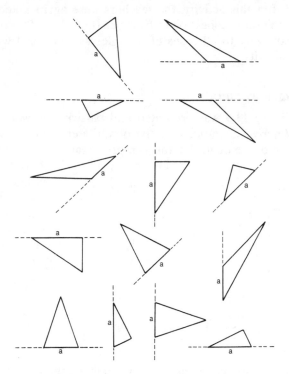

Fig. 19.7. The triangles given in the altitude task

For homework, teachers were asked to create an activity to develop the altitude concept, using the trianquad exercise as a model and the student mistakes as the basis for the relevant nonexamples. An analysis of their homework showed that most of them constructed a balanced set of examples and nonexamples (the common mistakes) in a reasonable logical order.

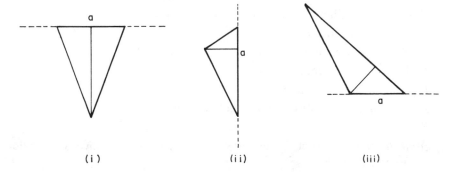

Fig. 19.8. "Altitudes" as drawn by one student

A month after this activity, the teachers were again asked to construct altitudes for the same collection of triangles (fig. 19.7). This time all twenty responded correctly to all types of triangles as compared with only eight prior to the activity.

Activity 2: Quadrilaterals

Stage 1. Figure 19.9 was presented and participants were requested to give three *different* definitions for the quadrilaterals in sets *A, B,* and *C.* Note that each of the quadrilaterals is representative of the infinite number of such figures in-cluded in each of the sets *A, B,* and *C.* As usual, the Venn dia-gram is used to show set inclusion (i.e., *A* ⊆ *B* ⊆ *C*).

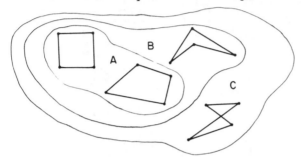

Fig. 19.9. Sets of quadrilaterals

Part of the discussion concerned with verbalizing the definition of quadri-laterals included only in set *A* follows.

T: "Quadrilaterals" with parallel sides . . .

R: ? ?

T: Only by chance . . .

R: ? ?

T: A closed figure with four sides.

R: Is this ⟋⟍ not a closed figure with four sides?

Teachers: Yes! There are four sides.

R: But it is not included in *A.*

Teachers: Concave and convex . . .

T: Four sides and four angles . . .

T: One angle less than 180° . . .

T: Four-sided polygons with no acute angles.

R: ? ?

Same teacher (correcting herself): . . . which do not have any angle greater then 180°.

T: What about this figure? Is it a polygon? ⟋⟍⟋

R: What is a polygon?

T: A closed figure whose sides are straight-line segments.

R: So, is it a polygon?

Teachers: Yes!

T: But a geometric figure must have an area, and what is the area of this figure?

R: ? ?

The participants concluded that whether this is a polygon or not depends on the definition.

T: To be safe we must add to the definition of A that opposite sides (unextended) should not intersect.

The discussion produced the following three definitions:

Set A: Four-sided polygons whose internal angles are less than 180° and whose opposite sides (unextended) do not intersect.

Set B: Four-sided polygons whose opposite sides (unextended) do not intersect.

Set C: Four-sided polygons (opposite sides may intersect).

In the process of verbalizing the definitions, participants explicitly remarked on the following:

1. The inclusion relationship between the sets
2. When a set becomes larger the definition includes fewer requirements (is less restrictive; has fewer critical attributes)
3. Some arbitrariness in definitions, yet from the moment a definition is adopted, it becomes the final criterion for deciding whether an example belongs or does not belong to the concept

Stage 2. Participants were requested to perform "concept analysis" of the quadrilateral concept according to the three definitions, starting with set *A*. They looked for additional examples, listed all the attributes of convex quadrilaterals, and in the process distinguished between critical and non-critical attributes.

The search for critical attributes in sets *B* and *C* created an opportunity for the teachers to overcome some of their own misconceptions. For instance, one of the attributes, which was suggested for the quadrilaterals of set *A*, was that each has two diagonals. While examining this attribute in set *B*, one of the participants exclaimed, "Some quadrilaterals here have only one diagonal!" Her concept image (like that of many of the participants) included only diagonals that are "inside" the figure. A discussion about diagonals, and a definition followed by examples and nonexamples, enabled her and her colleagues to "discover" by themselves the diagonals in the figures of set *C*.

Another surprise occurred when the participants discussed the sum of the internal angles of the quadrilaterals. This sum was easily seen to be 360° for the figures in sets *A* and *B*. But what about set *C*? It was not easy to accept that here the sum is less than 360°. Many teachers wanted to include the pair of opposite angles. This led to a careful rephrasing of the definition for set *C*, in a form somewhat like that given by Galbraith (1981):

> A quadrilateral is what you get if you take 4 points *A, B, C, D* in a plane and join them with the straight lines *AB, BC, CD, DA!*

It is then clear which are the internal angles in this figure. At this point, an inevitable discussion arose as to whether or not it is worth considering these figures as quadrilaterals in teaching geometry in elementary school. Participants now appreciated why many elementary school textbooks adopt the set *B* definition as "the definition" of quadrilaterals.

Stage 3. The participants were informed about some research results on this topic (Hershkowitz and Vinner 1983). This included a description of how *student* concept images of quadrilaterals change with class grade, from squares only to all convex quadrilaterals, and then also to concave quadrilaterals.

In this research, the students were requested to explain why they concluded that a shape was not a quadrilateral. A few such explanations were presented to the teachers and analyzed in a group discussion.

Galia (grade 5):   This figure does not look like a quadrilateral.

Johnny (grade 6):   This is not a quadrilateral because it is two triangles combined.

Miriam (grade 7):   This is not a quadrilateral because it has six sides and six angles.

Alex (grade 6):   This is not a quadrilateral because it does not have four equal sides and four equal angles.

The teachers paid attention to the different levels of reasoning. Galia and Johnny apparently judged the figure by its global appearance (van Hiele level 0), whereas Miriam and Alex related to the figure's attributes (van Hiele level 1). Miriam related to a critical attribute of the quadrilateral concept (four sides, four angles). She rejected the figure as a concept example because she could not find in it these attributes. Alex related the noncritical attributes (the attributes of a square: equal sides and equal angles), as if they were critical attributes (i.e., necessary to each concept example).

This discussion was followed by a worksheet containing twenty-four different student explanations. Participants, working in pairs, were asked to evaluate and analyze the explanations. In the process, there was clear evidence of their growing appreciation of the role of different attributes in the formation of the concept image and, even more, how some noncritical dominant attributes lead to various misconceptions (see Alex's explanation).

These two activities exemplify a model for teacher in-service training, made up of a number of important components. There is, of course, the geometric knowledge component in the sense that after these activities the teacher's conception of the concepts discussed is much improved. But because this is done at an advanced didactical level, teachers are led to pay attention to metaconcepts like definition, examples, and nonexamples, as well as to be aware of and analyze their students' reactions. We believe that this kind of activity can contribute to breaking the vicious cycle of teachers' and students' misconceptions in geometry.

REFERENCES

Galbraith, P. L. "Aspects of Proving: A Clinical Investigation of Process." *Educational Studies in Mathematics* 12 (February 1981): 1–28.

Herron, Dudley J., E. Kundayo Agbebi, Larry Cattrell, and Thomas W. Sills. "Concept Formation as a Function of Instructional Procedure." *Science Education* 60 (July-Sept. 1976): 375–88.

Hershkowitz, Rina, and Shlomo Vinner. "The Role of Critical and Noncritical Attributes in the Concept Image of Geometrical Concepts." In *Proceedings of the Seventh International Conference for the Psychology of Mathematics Education,* edited by Rina Hershkowitz, pp. 223–28. Rehovot, Israel: Weizmann Institute of Science, 1983.

———. "Children's Concepts in Elementary Geometry: A Reflection of Teachers' Concepts?" In *Proceedings of the Eighth International Conference for the Psychology of Mathematics Education,* edited by Beth Southwell, Roger Eyland, Martin Cooper, John Conroy, and Kevin Collis, pp. 63–69. Darlinghurst, Australia: Mathematical Association of New South Wales, 1984.

Hoffer, Alan. "Van Hiele–Based Research." In *Acquisition of Mathematics Concepts and Processes,* edited by Richard Lesh and Marsha Landau, pp. 205–27. New York: Academic Press, 1983.

Usiskin, Zalman. *Van Hiele Levels and Achievement in Secondary School Geometry.* Department of Education, University of Chicago. (ERIC Document Reproduction Service No. SE 038813), 1982.

Vinner, Shlomo, and Rina Hershkowitz. "Concept Images and Common Cognitive Paths in the Development of Some Simple Geometrical Concepts." In *Proceedings of the Fourth International Conference for the Psychology of Mathematics Education,* edited by Robert Karplus, pp. 177–84. Berkeley: Lawrence Hall of Science, University of California, 1980.

20

Geometry for
Secondary School Teachers

Margaret A. Farrell

Teachers of secondary school geometry, like all teachers, must be concerned about the answers to two questions: (1) How can the central characteristics of the discipline be translated in ways that students understand but that do not distort the nature of the discipline? (2) What special characteristics of students may hinder or enhance their understanding of the subject? Secondary school geometry highlights axiomatics, novel problem solving, and discrimination of spatial relationships. The geometry course is typically taught to 14-to-15-year-old students, many of whom still depend on concrete approaches to problems. This article outlines some sample strategies for college geometry courses that reflect these two concerns.

The Nature of Geometry

For the teacher of secondary school geometry, the question "What is geometry?" is a question about inherent structure and the view of that structure that should be communicated to secondary school students. More than fifty years ago, Veblen (1934) spoke of geometry as being both a branch of mathematics *and* a branch of physics. Both branches, he believed, should be treated in a high school course. Further reading of Veblen shows that he considered the mathematical part of geometry to be the axiomatic structure and its content of ideal points, whereas the part that belonged to physics included intuitive-experimental tasks. Veblen's view reflects both the historical development of geometry and contemporary uses of geometry. As a dynamic discipline, geometry can be applied to theoretical or real-world problems—the products aspect. At the same time, geometry can be used to extend one's knowledge and understanding of both the ideal world of mathematics and the real world in which we live—the processes aspect. The product and process aspects are actually a pair of inseparable constructs that are considered separately in order to understand the discipline better (Farrell 1967).

The products of geometry. Secondary school geometry would fail to be true to its heritage if it did not emphasize the understanding of a mathematical system. What are the differences between assumptions and theorems? Do future teachers realize that some geometry textbooks include as assumptions statements that *can* be proved? Do they understand why the authors of such texts made these choices? The college instructor has an opportunity to explore the raison d'être for *assumptions* in an axiomatic system, the characteristics of *well-formed assumptions,* and the historical evolution of the terms *axiom, postulate,* and *assumption.* Many college students complete a mathematics major without an appreciation of the inventive nature of mathematical definitions. On an intellectual level, these students will agree that definitions can be constructed; on an emotional level, they behave as if the substance and form of definitions in geometry were fixed and inevitable.

A schematic attempt to illustrate the hierarchical relationship among these separate products is found in the circle on the left side of figure 20.1.

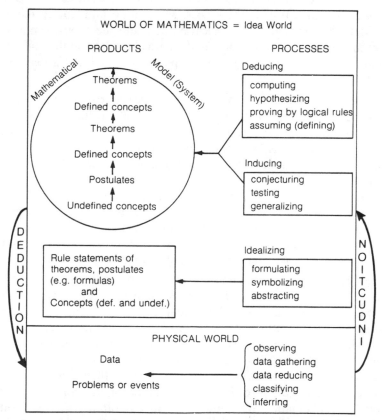

Fig. 20.1. A model of mathematics. Reprinted, with permission, from Farrell and Farmer (1980).

The term *postulate,* rather than *assumption,* is used in the diagram because historically *postulate* referred to statements with geometric context. The importance of defined concepts to extend the deductive chain further is emphasized by the repetition of the term to indicate the need for the construction of additional definitions if the chain is to be continued. The interconnected and deductive nature of the system is illustrated by including all these products in a circle. Yet the same terms are listed in the rectangular shape to suggest that many of the individual concepts and theorems are also studied for themselves and their properties rather than solely to advance the deductive chain.

Finally, the observables from the real world are listed as *data* and *problems or events.* In spite of the small area in which these last products are found, their importance in the development, applications, and learning of geometry is tremendous. Physical models, such as sugar cubes, shoe boxes, windows, and frames of all kinds, suggest properties of one or more geometric models. The appropriate use of such physical models helps students to visualize, find patterns, and observe differences. More important, the recognition of the differences between the physical models and the thought models they represent clarifies the nature of a mathematical model and the inability ever to find a one-to-one match between a physical model and a mathematical model.

The processes of geometry. Most mathematics courses seem to emphasize the products of mathematics, and yet, learning about the processes may be more important to most secondary school students. In figure 20.1, each of the sets of products is linked to sets of processes. The two-pronged set connected to the mathematical system is meant to convey the notion that in developing (or learning about) an axiomatic system, the student moves back and forth between inductive processes and deductive processes. Diagrams, hatch marks, and abbreviated notes may appear on paper as a problem-solver considers relationships and tests ideas. Notice that processes such as *testing* and *conjecturing* suggest some hesitant starts and some trial and error, as well as a systematic problem-solving approach.

The three processes listed under *idealizing* are critical to the understanding of geometry. Consider the following sequence of examples: a part of the superstructure of a bridge, a triangular picture of a part of the bridge superstructure, an outline of the same triangular shape labeled *ABC,* the *actual* triangle (i.e., the thought model), a written phrase about the thought model such as $\triangle ABC$. From the real world to the idea world, the student must use *abstracting* at more than one level and, finally, writes about the thinking being done by *symbolizing* the name of the concept. From the history of mathematics, we know that artisans often *formulated* rules or principles before anyone proved that the rules followed logically from a deductive system. However, the process of *formulating* is important for more than historical reasons. When a theorem such as *The altitude to the hypotenuse of a right triangle is the mean proportional between the segments of the hypotenuse* is studied in a secondary school geometry class, it must be considered in terms of its place in the axiomatic chain and in terms of its applications. The first role deals with the ways in which the theorem becomes part of the deductive chain. How can it be proved and how can it be used to extend the deductive chain? The second role deals with the formula-aspect of the theorem. The statement of the theorem represents a rule or formula that can be applied to answer quantitative questions.

The processes listed in the Physical World portion of the diagram are probably those that Veblen would have characterized under the physics aspect of geometry. From the point of view of problem solving, these processes are critical in all areas of mathematics. Their particular application in geometry will involve intellectual skills frequently related to spatial visualization. Finally, the curved arrows between the Idea World and the Physical World are meant to suggest other ways in which inductive and deductive reasoning are used. Set designers in theaters make constant use of geometric concepts to create special effects, and one of the most famous contemporary examples of the search for a mathematical model is the history of Watson and Crick's decision to use the double helix as a thought model of DNA.

The Nature of the Students

There are three student variables of particular interest to geometry teachers: cognitive development, spatial abilities, and attitudes toward geometry.

Cognitive development. What cognitive abilities are needed by students in a secondary school geometry class? They have to be able to hypothesize, reason deductively, understand the role of mathematical models, and understand the difference between defining and deducing (Farrell 1967; Farrell and Farmer 1979). All these abilities are characteristics of Piaget's formal operational stage. Tests of students in geometry classes by various cognitive development measures show that a minimum of 30 percent of these students reason at the concrete operational level, with another 30 percent to 40 percent being assessed as transitional reasoners (Farrell and Farmer 1985; McDonald 1982). The results of these recent assessments are in agreement with those summarized by Neimark (1975) and found for a similar population in Great Britain by Shayer and Adey (1981).

What are the first steps in adapting instruction and curriculum to the assumed cognitive developmental abilities in a geometry class? First, it is important to identify the *capabilities* of concrete operational students, rather than concentrate on their limitations. For example, they *can* reason inductively and learn some material if it is couched in concrete or familiar terms.

It is particularly important to realize that transitional students are in a developmental hiatus and need the stimulation of activities that vary gradually from concrete to abstract. Researchers, such as Pluta (1980), who studied the learning of college students in a unit on groups, found that even the formal student profited from initial concrete, inductive approaches (as opposed to less interactive, deductive approaches) to abstract material. From a Piagetian point of view, instruction likely to promote meaningful learning for all students would have the following characteristics: the use of concrete models, emphasis on inductive reasoning, and strategies that promote the interaction of students with the teacher and each other. Throughout this article, sample activities with one or more of these characteristics will be outlined.

Spatial visualization. Geometry, whether restricted to traditional *plane geometry* or couched in *transformational* terms, requires discrimination ability with visual material of varying degrees of complexity. Students have to be able to locate the *hidden,* or *turned,* shapes in order to solve even fairly straightforward congruence problems. When they must match overlapping triangles that are similar, rather than congruent, the task becomes even more difficult (see fig. 20.2).

In the diagram, $\overline{SP} \perp \overline{PQ}$; $\overline{PQ} \perp \overline{QT}$; $\overline{QR} \perp \overline{PR}$. Show that PQ is the mean proportional between QT and PS.

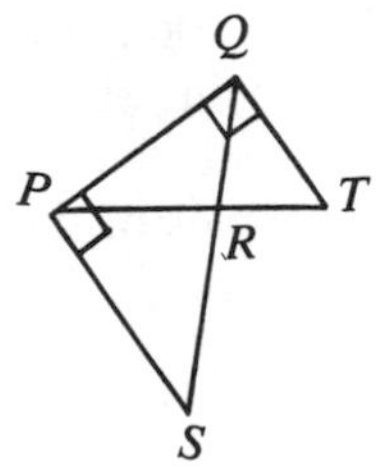

Fig. 20.2

The particular form of spatial ability embedded in problems like the one in figure 20.2 was studied by Ranucci (1952), who subsequently devoted much of his writing in geometry to activities that gave students from elementary school to graduate school practice in the mental transformation of shapes.

Attitudes toward geometry. There is a historical record of the negative, anxious attitude of many students toward geometry (NCTM 1930). When college students with such an attitude are asked to trace the origin of their anxiety, they will usually identify one or more of the following: too much memorization, little or no attempt to relate the subject to the real world, and the rigid attitude of the teacher toward alternative approaches (Buerk 1982).

Unfortunately, these reasons mirror the way too many teachers have interpreted the discipline of geometry. In the name of rigor, rigidity occurs. Euclid's hard-won deductive structure is presented as if there had been no human errors or long-winded proofs. Preservice teachers of mathematics are aware of this dislike of geometry by some of their nonteaching friends and are often fearful that their first teaching assignment may include this course. Neither the college instructor nor the secondary school teacher can dismiss the effect of negative attitudes and anxiety on the learning of geometry. For the college teacher of preservice teachers, the first task is to help these students face their fears, understand the attitudes of others (and perhaps their own), and learn a different view of the discipline of geometry.

Sample Course Strategies and Materials

In the rest of this article, sample problems, classroom activities, and projects are presented in the context in which they have been successfully used with college mathematics majors, most of whom were preservice mathematics teachers. Many of the strategies and activities have also been used in in-service courses for secondary teachers who, in turn, have used them with their own students.

The Dynamics of Group Problem Solving

Geometry, with its rich tradition of classical problems and contemporary usefulness in terms of mathematical models, seems to be especially suited for problem-solving activities. It appears that the understanding of geometry is deepened as college students interact with each other in analyzing constructions, devising proofs, or struggling to find a best-fit geometric model for a problem situation. Yet, as noted earlier, fear of the content can be a deterrent to successful problem solving. Thus, at the beginning of a course, problem-solving activities should have a high potential of success for most students. Early exercises can be based on secondary school geometry so that students have a chance to recall, apply, and strengthen their secondary school background. Two such exercises follow:

Exercise 1. Construct triangle *ABC,* given side *a, h_a* (altitude to side *a*), and *m_a* (median to side *a*).

Exercise 2. The base of a triangle is 12 cm, and the median to that base is 5 cm. The area of the triangle is 24 cm². Find the lengths of the other two sides.

Even these relatively simple exercises can frighten an otherwise competent mathematics major who has not had to recall and apply most of the geometric building blocks learned in high school in a single college mathematics course. In fact, every construction problem, other than the standard set (e.g., bisect an angle), seems to be difficult for most students, primarily because they have not been taught a method of analysis.

Exercise 1, therefore, offers the instructor an opportunity to introduce a method of analysis by way of some probing questions:

● How did you draw triangle ABC?

- What is the effect of drawing an isosceles triangle?
- In this exercise, does a consideration of *cases* make it easier to explain the solution?
- If so, describe those cases. . . .

After a series of questions such as these, the instructor and the class can discuss the meanings of *determined, overdetermined,* and *underdetermined* and summarize a general strategy associated with the analysis of construction and locus problems:

a. Sketch carefully the *desired result*. Label and mark (tick marks, shading, . . .) the given parts. (For Exercise 1, the student might sketch the diagram in fig. 20.3.)

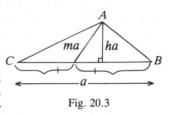

Fig. 20.3

b. Study the resulting diagram and mark or symbolize *additional* relationships (e.g., right angles when given a circle and a tangent line). (Now the student adds right angle marks and tick marks for equal lengths.)

c. Locate one part (segment, triangle, . . .) that *can* be constructed and then complete the construction. (The student reduces the original problem to "What triangle is determined?")

d. Compare the completed portion of the construction with the sketch and then repeat steps *b* and *c* until the construction has been completed.

A follow-up set of exercises could include a novel dissection problem (Exercise 3) and a locus construction problem (Exercise 4). Exercise 4 provides the teacher with an excellent opportunity to discuss cases. For instance, what will happen if $\overleftrightarrow{CD}$ is parallel to $\overleftrightarrow{AB}$?

Exercise 3.   Divide a parallelogram into three equal parts by drawing lines through one vertex.

Exercise 4.   Construct a circle that passes through points *C* and *D* and is tangent to $\overleftrightarrow{AB}$. (The student may have to be taught how to construct a mean proportional to solve this problem.)

Although any of these exercises could be assigned as outside work, there is enormous value in using them as the context of group problem-solving activity. In mathematics classes at the college level, however, students have generally been listeners and note-takers. All the factors related to negative attitudes toward geometry simply heighten their concern at the risk involved in small-group work. Thus, it is important that the perceived risk be minimized as much as possible. In addition to a careful choice of content, especially at the early stages, the instructor needs to allow sufficient time for "playing around with the data." Depending on the problem, the class, and the instructor's style, the instructor or a student might complete the solution at the chalkboard. In any event, it is vital that the instructor, with the class's help, summarize important strategies, discuss false leads, and highlight special instances.

As the students become more comfortable with the problem-solving sessions and more adept at retrieving useful information from resources, the problems can become more challenging. They can be started in class and left to the student to complete outside class (individually or with members of the class). Solutions and strategies should be summarized increasingly by student-solvers, rather than by the instructor. Sample examples of various degrees of difficulty are included next.

> A circular brick chimney is 3 feet in diameter and 45 feet high. A spider at the bottom of the chimney makes ten circular trips to catch a fly that is at the top of the chimney directly above the spider. How long is the helix, or path of the spider?
>
> Two towns, A and B, are on the same side of a river. A bridge is to be built across the river so that the sum of the distances from A and B to the entrance of the bridge is a minimum. Where should the bridge be located?
>
> With a given radius, construct a circle that is orthogonal to a given circle and tangent to a given line.
>
> A penny rolls outside triangle ABC from A to B to C. What is the locus of its center; the length of the path of the center, given the sides of the triangle (a, b, c); and the radius of the penny (p)?

The Four Freedoms

Geometry teachers, perhaps more than teachers of any other branch of mathematics, have a difficult time reconciling the desire for rigor and exactness of language with a teaching approach that illuminates the problem-solving nature of the discipline. Students' attempts to paraphrase a principle or to suggest a method longer than, or different from, that in the text are often interpreted as sloppy mathematics. The most sacrosanct of all rules seems to be that an argument is not a proof unless it is written in two columns.

The late Herbert Fremont (1969) defined four freedoms that should apply to the learning of mathematics: (1) freedom to make a mistake, (2) freedom to think for oneself, (3) freedom to ask a question, and (4) freedom to choose methods of solution. The instructor can structure lessons and exercises to illustrate a belief in these four freedoms while teaching college mathematics majors more about the nature of geometry. One way of doing this is to "think aloud" as a proof is analyzed. In the sample that follows, the proof of the Euler line theorem is approached in this way.

The Euler line theorem: *The circumcenter, centroid, and orthocenter of any triangle are collinear, and the distance from the orthocenter to the centroid is twice the distance from the centroid to the circumcenter.*

| **Talking aloud** | **Chalkboard work** |

I'll draw *any* triangle. Better make sure it doesn't look isosceles. I'm going to sketch in two perpendicular bisectors (to find a reasonable location for the circumcenter) [marks the intersection *S*] . . .

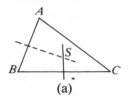

(a)

and then erase them. I may not need them all, and they're liable to be distracting.

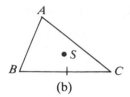

(b)

I'll locate the centroid in the same way. [Sketches two medians. Marks *G* and

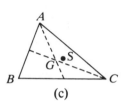

(c)

erases the medians.] I may need one or more of these later, but let's wait and see. Now I'll locate the orthocenter. [Marks it *H*.] Let's write down what we're looking for and study the diagram.

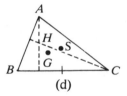

(d)

Typical college students will copy what the instructor actually finishes with on the chalkboard but can't help but learn an implicit lesson about (1) the value of careful sketches, (2) the need to keep the diagram uncluttered, and (3) behavior that suggests that it's important to try ideas that might not be correct or immediately useful.

We've got to show that *S*, *G*, and *H* are collinear and that $\overline{SG} = \frac{1}{2}(\overline{GH})$. That 1:2 ratio should ring a bell. Let's brainstorm and think of other 1:2 relationships. [Students suggest midpoint, medial triangle, and the division of a median by the centroid.] Well, since we're working with a centroid, let's put in a median. [Draws $\overline{AA'}$.]

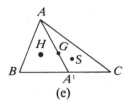

(e)

[Stands back and looks at the diagram.] Sure seems as if we should draw $\overline{SG}$ and $\overline{SA'}$, doesn't it? Now if we extend $\overline{SG}$, we'd have vertical angles. Unfortunately, we can't assume that H is on line $\overleftrightarrow{SG}$. [Marks ratio on median and marks vertical angles.]

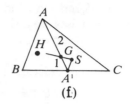

(f)

The instructor continues in this vein, using cues and developmental questions, to get students to suggest that $\overline{SG}$ be extended to some point, P, so that $GP = 2SG$. The final segment is drawn, and the students agree that P must be shown to coincide with H and discuss how that can be done.

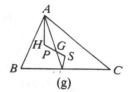

(g)

The instructor should probably continue analyzing the proof in this fashion if this is an early proof activity. The idea of the difference between "proof doing" and "proof writing" may begin to be considered by posing questions such as the following:

- Have I, with your help, just *done* a proof? *Yes.*
- Have I *written* a proof that you would judge satisfactory? (Students may not want to seem to criticize the instructor.) *Not really.*
- If I gave you a pop quiz right now and asked you to write a proof for this problem, would anybody here not be sure of a perfect paper? (Unless the class is unusual, that question is bound to result in verbal and nonverbal expressions of concern.)

A good assignment for the class is to write this proof with half the class doing the traditional two-column proof and the rest writing a paragraph style of proof. A satisfactory sample of each format could be distributed later, and issues of rigor, style, and acceptable variations and shortcuts could be discussed. Once again, the instructor will have the opportunity to help these students distinguish between doing a proof and writing a proof. The first requires the freedom to think and try ideas and is characterized by a somewhat erratic movement back and forth between inductive and deductive reasoning. Secondary school students who can *do* a proof but cannot yet *write* a satisfactory proof need to know that they have good problem-solving skills, even though they must work at the mechanics of proof writing. Prospective teachers need to learn that during proof doing, rigor and precision take a back seat to the generation of ideas.

Proof doing is a form of problem solving with its own set of heuristics. The instructor can help students realize that key strands or themes connect sets of proofs and that successful proof doers study the diagrams and the given information to identify possible key strands. Suppose the following three exercises are assigned soon after work on the Euler line theorem. The class of preservice teachers could be asked not to do the proofs at first but to study the entire set, identify possible common relationships, and write (or in groups, discuss) a range of approaches.

- Given: *ABCDE* is drawn so that $\overline{AE} \parallel \overline{CD}$ and *AE = CD; P, K,* and *M* are midpoints of $\overline{AB}$, $\overline{BC}$, and $\overline{ED}$, respectively.
  Prove that $\overline{KE}$ bisects $\overline{PM}$.

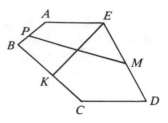

- Prove that the centroid of a given triangle is also the centroid of its medial triangle.
- Prove that each median of the given triangle is bisected by a side of the medial triangle.

Another strategy to drive home the fact that problem solving of any substance requires germinating time is to assign exercises that are not due for two, three, or more weeks. Here are a few that can be used in this way:

- Prove that the sum of the perpendiculars from any point inside an equilateral triangle to its sides is constant and equal to the length of any altitude of the triangle. (*Hint:* Begin by drawing one altitude and a line from the given internal point parallel to a side.)
- *D, E,* and *F* are the midpoints of the sides, *BC, CA, AB,* respectively, of a triangle, *ABC. FG* is drawn parallel to *BE* meeting *DE* produced in *G*. Prove that the sides of the triangle *CFG* are equal to the medians of the triangle *ABC*.
- If an isosceles triangle *PAB*, with base angles whose measures are 15°, is drawn inside square *ABCD*, then the points *P, C,* and *D* are vertices of an equilateral triangle.

Finally, proof doing can be enhanced by occasional problems that have numerical characteristics. The following examples can be used to help students hone their algebraic/geometric skills.

- Find the area of the square inscribed in an isosceles triangle whose base is 60 cm long and whose equal sides are each 50 cm long. Let two vertices of the square lie on the base of the triangle.
- The hypotenuse of a right triangle is 10 cm, and the radius of the inscribed circle is 1 cm. Find the perimeter of the triangle.
- The interior of a box is a cube with an edge of 1 foot. What is the volume of the largest sphere it will hold?
- An arch is built in the form of an arc of a circle and is subtended by a chord about 30 feet long. If a chord about 17 feet long subtends half that arc, what is the radius of the circle?

The point was made that proof doing is not the same as proof writing and that both require instruction. Why is the writing of two-column proofs so difficult for secondary students? (See Senk [1985] for results on achievement on proof-writing measures.) An analysis of the logical form that is used most often may help to explain the problem. A typical proof relies on the repeated use of implications, especially *modus ponens:*

$$[(p \to q) \wedge p] \Rightarrow q$$

In a standard geometric problem, this symbolic representation is applied in a complex fashion. Consider the sequence of statements that follows.

Statements	Reasons
1. $\triangle ABC \cong \triangle DEF$	1. Given.
2. $\overline{AB} \cong \overline{DE}$	2. If two triangles are congruent, their corresponding sides are congruent.

Now let's see which parts of this miniproof match the propositions in modus ponens. The reason *If two triangles are congruent, their corresponding sides are congruent* corresponds to $p \to q$. (Of course, we mean "If any two. . . ." Thus, the antecedent is universally quantified.) The statement $\triangle ABC \cong \triangle DEF$ is represented by p. (Here is a case of a specific pair of triangles that are congruent.) Therefore, we can conclude that $\overline{AB} \cong \overline{DE}$, or q. However, the student begins the two-column proof by writing p for the special case, then derives q for this same special case, and finally lists as a reason for writing q the general implication, $p \to q$. No wonder students fail to see why the general principle is called a *reason* and think intuitively that the written sequence is somehow unreasonable. When a proof is analyzed, our thoughts are more likely to correspond to the following sequence.

> *Let's see. $\triangle ABC$ is given congruent to $\triangle DEF$. I know that congruent triangles have corresponding sides congruent. So $\overline{AB}$ must be congruent to $\overline{DE}$.*

If we consider the typical cognitive development of students in secondary geometry classes and the awkward complexity of the two-column proof format, it is easy to understand why many students seem to find geometry hard and why some succeed only by memorizing. An excellent topic for a college geometry lesson is "What Is a Proof?" when an analysis of proof forms and the logical underpinnings of proof might be discussed. The discussion in a college class might well hinge on the relative emphasis to be given to proof doing and proof writing and to problems with real-world characteristics versus those that seem removed from the physical world.

Productive Thinking

The consideration of geometries such as taxicab geometry (Martin 1975) provides students with excellent real-world applications of definitions and postulates. The activities that are suggested in the next section are additional ways to help students understand and feel more confident about geometry.

Behaving like a geometer. Geometry teachers could use some of the ideas developed by Harold Fawcett (1938) with their secondary school students. Fawcett described an entire secondary school geometry course in which students behaved like geometers and constructed their own definitions. College mathematics majors may need a different approach, since they already have preconceived notions of geometry (not necessarily good) and some background knowledge (not necessarily bad).

An instructor can introduce students to the idea of inventing their own definitions by sharing the beginnings of such a system with them. Farrell (1970) outlined an approach to area in which the unit of area would be the region enclosed by an equilateral triangle rather than a square. The coordinate system associated with this unit is one whose axes intersect at an angle of 60°. The following two exercises, which require students to develop theorems based on other sets of nontraditional definitions, can be used to clarify the meaning of axiomatics.

1. Assume the following two definitions:
   DEFINITION 1: *A kite is a quadrilateral such that exactly two pairs of adjacent sides are congruent.*
   DEFINITION 2: *Two kites are congruent if each pair of corresponding sides are congruent and each pair of corresponding angles are congruent.*

   Write a chain of theorems and corollaries based on the definitions above. Be sure to consider at least a few special cases of kites (include definitions for these special kites). Each statement in the chain should be as useful as possible. The chain may include statements on properties of kites as well as ways to prove two kites are congruent. Defend the validity of each statement in the chain (i.e., each theorem and corollary) by a short paragraph.

2. A. Suppose that a chain of theorems begins with the following postulate: *The area of a right triangle is equal to one-half of the product of its legs.* Use the postulate to prove theorem 1 in the chain.

   THEOREM 1: *The area of any triangle is equal to one-half the product of a base and the altitude to that base* (two cases: acute triangle and obtuse triangle).

   B. The next theorems in the chain are the following:
   THEOREM 2: *The area of a parallelogram is equal to the product of its base and altitude.*

   THEOREM 3: *The area of a trapezoid is equal to one-half the product of its altitude and the sum of its bases.*

   THEOREM 4: *The area of a regular polygon is equal to one-half the product of its perimeter and its apothem.*

Draw a tree diagram that describes the axiomatic chain consisting of the postulate and the four theorems.

C. *If, in a new area chain, we began by postulating that the area of a regular polygon is equal to one-half the product of its perimeter and its apothem, what theorem might we state as theorem 1? Explain your answer.*

Class discussion of the results can lead to a better understanding of choices to be made in designing a geometric system and, perhaps, more insight into the famous history of the Fifth Postulate. Which answers are correct, more meaningful, more useful? It will soon become clear to the class that until some ground rules are developed and accepted, judgments on the quality of the answers cannot be completed. The exercises and discussions have the dual characteristics of reflecting both the process and the product aspects of geometry while engaging students in a constructive learning activity.

Writing about geometric ideas. A teacher of geometry must be more than a good problem solver. To communicate geometry to others, the teacher must be able to talk about geometry in language other than that which mimics the textbook. Writing offers this opportunity, with the advantage that the writer has a chance to edit and improve the product. However, writing is more demanding than speaking, since the listener helps the speaker with verbal or nonverbal cues. In addition to paragraph proofs, the instructor can assign exercises in which the students must answer questions of why? when? how? or write a creative essay:

1. Are the nine points of the nine-point circle ever reduced to 8 (or 5, or fewer than 5) distinct points? When and why?

2. Choose *one* of the following sentences as the topic sentence of a 75-to-100-word paragraph. Use as much imagination as possible and insert correct mathematical concepts and relationships. Include at least five major concepts and illustrate or extend these concepts as part of the "story line."

    Topic sentence *a:* I am a circle in a coaxial family.

    Topic sentence *b:* I am a point on a radial line.

    Topic sentence *c:* I was an ideal point on an ordinary line.

Students can also be asked to write a paper on some famous geometer or to tell the story of the Fifth Postulate as an adventure. Such exercises serve the double benefit of encouraging mathematics students to *read* about geometry and those who developed it and giving these future students practice in one form of communication.

Summary

The activities suggested in this article are designed on the basis of the two concerns cited at the beginning of the article: the nature of geometry, and student variables. It is not surprising that in its latest set of guidelines for the preparation of mathematics teachers, the Mathematical Association of America (1983) reiterated these same concerns. Nor is it surprising that recent reports on the quality of liberal education for all college students identify the same need for problem-solving activities, writing and reading assignments that stress analytical thinking, and content that emphasizes the structure of, and human developments in, the disciplines.

REFERENCES

Buerk, D. "An Experience with Some Able Women Who Avoid Mathematics." *For the Learning of Mathematics* 3 (November 1982): 19–24.

Farrell, Margaret A. "Area from a Triangular Point of View." *Mathematics Teacher* 63 (January 1970): 18–21.

_____. "Pattern Centering and Its Relation to Secondary School Geometry Teaching: The Formation of Hypotheses" (Doctoral dissertation, Indiana University, 1967.)

Farrell, Margaret A., and Walter A. Farmer. "Adolescents' Performance on a Sequence of Proportional Reasoning Tasks." *Journal of Research in Science Teaching* 22 (September 1985): 503–18.

_____. *Systematic Instruction in Mathematics for the Middle and High School Years.* Reading, Mass.: Addison-Wesley, 1979.

Fawcett, Harold P. *The Nature of Proof.* Thirteenth Yearbook of the National Council of Teachers of Mathematics. New York: Columbia University, Teachers College, 1938.

Fremont, Herbert. *How to Teach Mathematics in Secondary Schools.* Philadelphia: W. B. Saunders Co., 1969.

Mathematical Association of America. *Recommendations on the Mathematical Preparation of Teachers.* MAA Notes No. 2. Washington, D.C.: The Association, 1983.

Martin, George E. *The Foundations of Geometry and the Non-Euclidean Plane.* New York: Springer-Verlag, 1975.

McDonald, Janet L. "The Role of Cognitive Stage in the Development of Cognitive Structures of Geometric Content in the Adolescent" (Doctoral dissertation, State University of New York at Albany, 1982). *Dissertation Abstracts International* 43(1982): 733A.

National Council of Teachers of Mathematics. *The Teaching of Geometry.* Fifth Yearbook. New York: Columbia University, Teachers College, 1930.

Neimark, Edith. "Intellectual Development during Adolescence." In *Review of Child Development Research,* edited by F. D. Horowitz. Chicago: University of Chicago Press, 1975.

Pluta, Raymond F. "The Effect of Laboratory and Quasi-Lecture Modes of Instruction on Mathematical Learning of Formal, Traditional, and Concrete Operational College Students" (Doctoral dissertation, State University of New York at Albany, 1980). *Dissertation Abstracts International* 41 (1980): 925A.

Ranucci, Ernest R. "Effect of the Study of Solid Geometry on Certain Aspects of Space Perception Abilities." Doctoral dissertation, Columbia University, 1952.

Senk, Sharon L. "How Well Do Students Write Geometry Proofs?" *Mathematics Teacher* 78 (September 1985): 448–56.

Shayer, Michael, and Philip Adey. *Towards a Science of Science Teaching.* London: Heinemann Educational Books, 1981.

Veblen, Oswald. "The Modern Approach to Elementary Geometry." *Rice Institute Pamphlet* 21 (October 1934): 209–21.